NEW JERSEY ASK4 SCIENCE TEST

Loris Jean Chen
Science Department Chair,
Eisenhower School, Wyckoff, NJ

Lauren M. Filipek
Fourth Grade Teacher,
Freehold Learning Center, Freehold, NJ

BARRON'S

DEDICATION

Loris dedicates this book to her patient husband Jim and to the teachers who inspired her passion for science.

Lauren dedicates this book to the student in each of us, that we may never lose our love for learning, and to her wonderful family who exemplify unconditional love.

All inquiries should be addressed to:
Barron's Educational Series, Inc.
250 Wireless Boulevard
Hauppauge, NY 11788
www.barronseduc.com

ISBN-13: 978-0-7641-4303-8
ISBN-10: 0-7641-4303-4

Library of Congress Control Number: 2009930140

Date of Manufacture: November 2011
Manufactured by: B11R11, Robbinsville, NJ

PRINTED IN THE UNITED STATES OF AMERICA

9 8 7 6 5 4 3

10%
POST-CONSUMER
WASTE
Paper contains a minimum
of 10% post-consumer
waste (PCW). Paper used
in this book was derived
from certified, sustainable
forestlands.

CONTENTS

INTRODUCTION

In the spring, you will take an important test, the New Jersey Assessment of Skills and Knowledge of Science. This test measures your achievement in science. You have been studying for this test since Kindergarten. The questions are based on the big ideas of life science, physical science, earth science, and environmental science.

Some questions will measure your understanding of the use of tools, the procedures of experimental design, and the meaning of data or observations. Information for these questions may be displayed in pictures, tables, charts, or graphs. Other information may require you to read short passages of text.

Most of the test questions are in multiple-choice format. Some questions may require you to write a short answer. Practicing test-taking skills is one way to review information. This can help you be a little more confident during the actual test.

Be sure to get a good night's sleep before the test and eat breakfast the day of the test. These two things will help you stay focused.

During the test remember to read the questions carefully. If you do not know an answer go to the next question. Come back to the question if you have time at the end of the test. If you have time, go back and check your answers.

Each chapter of this practice book is organized into big ideas that can be tested. Each section has sample questions based on the big ideas. Answers to the questions are provided to help you understand why an answer is correct or incorrect. Finally, there are two complete tests with answers to help you learn how to answer questions in a timed setting.

TOOLS OF THE SCIENTIST

TOOLS OF THE SCIENTIST CONCEPT MAP

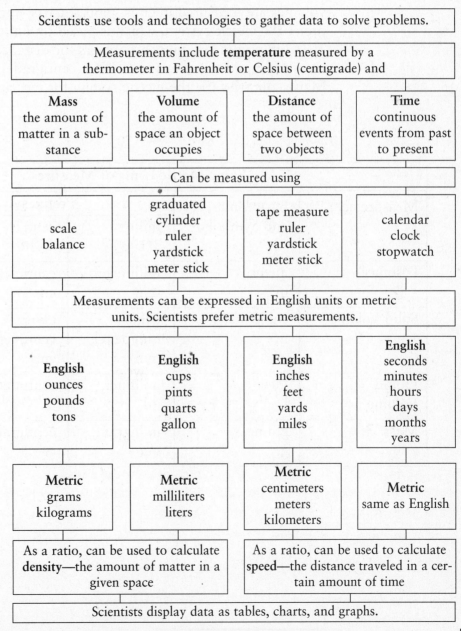

Scientists use tools and technologies to gather data to solve problems.

Measurements include **temperature** measured by a thermometer in Fahrenheit or Celsius (centigrade) and

Mass the amount of matter in a substance	**Volume** the amount of space an object occupies	**Distance** the amount of space between two objects	**Time** continuous events from past to present

Can be measured using

scale balance	graduated cylinder ruler yardstick meter stick	tape measure ruler yardstick meter stick	calendar clock stopwatch

Measurements can be expressed in English units or metric units. Scientists prefer metric measurements.

English ounces pounds tons	**English** cups pints quarts gallon	**English** inches feet yards miles	**English** seconds minutes hours days months years

Metric grams kilograms	**Metric** milliliters liters	**Metric** centimeters meters kilometers	**Metric** same as English

As a ratio, can be used to calculate **density**—the amount of matter in a given space	As a ratio, can be used to calculate **speed**—the distance traveled in a certain amount of time

Scientists display data as tables, charts, and graphs.

THE BIG IDEA

In order to gather data to solve a problem, scientists use measuring tools. The type of tool that a scientist uses depends on what is being measured. The unit of measurement depends what is being measured.

The information in this chapter is provided as background information. You will use the information and skills in the subject area chapters.

MAKING SENSE OF METRIC UNITS

The common units of measurement in the United States are based on the English system where weight is measured in pounds and ounces, distances are measured in inches, feet, yards, and miles, and volumes are measured in cups, pints, quarts, and gallons. Scientists prefer the metric system, which is based on units in multiples of 10. Common metric units are meters for distance, grams for mass, liters for liquid volume, and Celsius (centigrade) for temperature. Many countries use the metric system for measurements.

Some Metric Units of Measure				
Measurement	Basic Unit and Symbol	Prefixes		
		milli 1/1000	centi 1/100	kilo 1000
Distance	meter, m	millimeter, mm	centimeter, cm	kilometer, km
Solid Volume	cubic meter, m^3	cubic millimeter, mm^3	cubic centimeter, cm^3	N/A
Liquid Volume	liter, l	milliliter, ml	centiliter, cl	kiloliter, kl
Mass	gram, g	milligram, mg	centigram, cg	kilogram, kg

1.1 MEASURING MASS

Everything is either matter or energy. Objects are made of matter. The amount of matter in an object is called **mass**. Mass is often confused with weight, but they are different. Weight is the pull of gravity on the amount of matter in an object. Weight changes if gravity changes.

The moon's gravity is less than Earth's. When astronauts stood on the moon, their weights were 1/6 of their weights on Earth. Their masses were the same whether the astronauts were on the moon or on the Earth.

Mass can be measured using a scale or a balance. A digital scale for measuring mass is adjusted to take gravity into account. A balance compares an object of unknown mass to a known mass. Figure 1 shows a scale and a balance.

Figure 1

In the English system of measurement, weight is measured in ounces, pounds, and tons. In order to find the mass, measurements of weight must be adjusted to account for the pull of gravity. It is an awkward system for measuring mass.

The metric system is easier to use because it is a direct measurement of mass. Small amounts of mass are measured in grams. The symbol for gram is g. A large metal paper clip has a mass of about 1 g. Larger amounts are measured in kilograms. The symbol for kilogram is kg. A thousand paper clips would have a mass of 1 kg.

CHECK YOUR UNDERSTANDING 1.1 (For answer, see page 149.)

What is the mass of the cylindrical object on the balance in Figure 1?

Check your answer before going on to the next topic.

1.2 VOLUME OF A SOLID

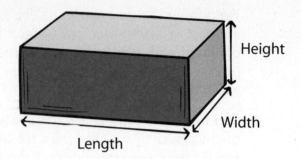

Figure 2

The box shown in Figure 2 is a rectangular box. How much space does the box occupy? To find the volume of the box, a scientist would measure the length, width, and height of the box. The volume can be calculated by multiplying the length by the width and by the height.

These dimensions are linear measurements or measurements taken in a straight line. Tools that could be used to collect data include a ruler, yardstick, or meter stick. The size of the box determines which tool to use.

If the box were small, a pencil box for example, a ruler would be a good tool to use. In the English measurement system, the length, width, and height would be recorded in inches and the volume in cubic inches. In the metric system, the length, width, and height would be measured in centimeters. The volume would be written as cubic centimeters (cm^3).

If the box were large, a toy chest for example, a yardstick or meter stick would be a good tool to use. In the English measurement system, the box dimensions would be measured in feet or yards and the volume recorded in cubic feet or cubic yards. In the metric system, the dimensions would be measured in meters. The volume would be written as cubic meters (m^3).

Did you notice that there is a superscript (raised) number 3 next to the unit of measure? This is because volume is measured by multiplying three dimensions.

CHECK YOUR UNDERSTANDING 1.2 (For answer, see page 148.)

A rectangular box has a length of 10 cm, a width of 5 cm, and height of 2 cm. What is the volume of the box?

Check your answer before going on to the next topic.

1.3 VOLUME OF A LIQUID

When water is in a cup, it takes the shape of the cup. When a cup of water tips over, the water will spread out over the surface of the table.

Water at room temperature is a liquid. Liquids take up a definite amount of space (volume), but they do not have a definite shape. Since the liquid substance does not have a constant length, width, or height, special tools are used to measure the volume of the liquid.

In the English measurement system, a liquid might be measured using a cup marked in ounces. Larger volumes include pints, quarts, and gallons. Milk is a common liquid packaged in quarts, half gallons, and gallons.

In the metric system, liquid volume is measured in liters. Soda is a common liquid sold in liters (l). Many bottled beverages have both the English and metric volumes listed on the label. Smaller amounts of a liquid are measured in milliliters (ml). There are 1,000 milliliters in a liter. It would take 237 milliliters to equal 1 cup or 8 fluid ounces.

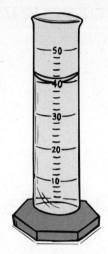

To measure a liquid volume, a scientist would use a graduated cylinder (see Figure 3). The cylinder has marks on the side similar to marks on a ruler. These marks are gradations, which is why the cylinder is said to be graduated. Notice that the liquid curves up the sides of the cylinder and flattens at the bottom of the curve. The volume is read at the bottom of the curve or meniscus.

**Graduated Cylinder
Figure 3**

CHECK YOUR UNDERSTANDING 1.3 (For answer, see page 149.)

How much water is in the graduated cylinder shown in Figure 3?

_____

Check your answer before going on to the next topic.

1.4 CALCULATING DENSITY

Mass ÷ Volume = Density

Density is the amount of matter in a given amount of space or mass per volume. Scientists use an object's mass and volume to calculate density by dividing the mass by the volume. Density is a property that can be used to identify the type of matter in an object.

If two objects appear to be of the same substance, density can be used to prove whether or not they are made of the same material. Differences in density can also be used to predict whether an object will sink of float in a fluid such as water or air.

The density of a pure substance does not depend on the size of the sample. The mass and volume change together in a unique ratio. Figure 4 shows two samples of the same substance. Even though sample A is larger than sample B, they have the same density because they are both the same substance.

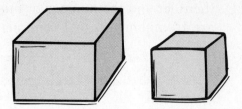

Sample A
Mass = 32 grams
Volume = 16 cubic centimeters

Sample B
Mass = 16 grams
Volume = 8 cubic centimeters

Figure 4

The calculator is a technology that a scientist could use to convert measurements of mass and volume to density. The unit of density is a combination of the unit used for mass and the unit used for volume, for example g/cm^3.

CHECK YOUR UNDERSTANDING 1.4 (For answer, see page 149.)

Calculate the density of the substance shown in Figure 4.

Check your answer before going on to the next topic.

1.5 MEASURING DISTANCE

Distance is the measurement of the space between two places or two objects. Small distances can be measured with a ruler. Longer distances can be measured with a yardstick or meter stick.

Think of a football game. In order to keep the ball, the team needs to advance 10 yards in four tries. If a play is very close, the referee will call for a measurement. To earn the first down, the ball must have advanced a distance of at least 10 yards. If the ball is inside the 10 yards, the other team will gain possession of the ball. The accuracy of the first down measurement can affect the outcome of the game.

Exact measurements are also important to scientists. Instead of using inches, feet, yards, or miles, scientists use metric units such as millimeters (mm), centimeters (cm), meters (m), and kilometers (km). Figure 5 shows the actual length of an inch on one side of the ruler and the number of millimeters on the other side. There are 10 millimeters in 1 centimeter.

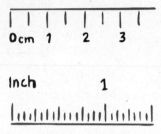

Figure 5

Figure 6 shows the relative length of a kilometer and a mile. This illustration is drawn to scale since a kilometer and a mile would be much larger than the paper!

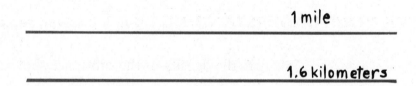

Figure 6

CHECK YOUR UNDERSTANDING 1.5 (For answer, see page 149.)

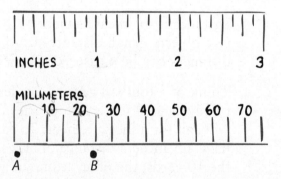

What is the distance between *A* and *B* in millimeters?

_____

Check your answer before going on to the next topic.

1.6 MEASURING TIME

Time has no dimensions. It moves in one direction and cannot rewind. Our system of time measurement is based on Earth's rotation around its axis (24 hours or 1 day) and orbit around the Sun (365¼ days or 1 year). The units of time are the same both in the English system and in the metric system because there are no dimensions of distance or mass.

Often scientists collect data that depends on how long an event lasts. Short time periods may be seconds or minutes. Very fast events may be measured in fractions of second called milliseconds. Clocks and watches are good tools for measuring short time periods, for example how long it takes for a sugar cube to dissolve in water.

Longer time periods could include hours, days, months, or years. A clock would be a good tool for timing events that take hours to complete. A calendar would be useful for tracking events that happen over days, months, or years, for example, the number of days it takes for the moon to complete one orbit of the Earth.

If an event needs accurate timing, a stopwatch might be used. A digital timer reports time as digits. An analog timer looks like a classroom clock with minute and second hands. (See Figure 7.)

Figure 7

CHECK YOUR UNDERSTANDING 1.6 (For answer, see page 149.)

What tool would you use to keep track of the passage of time between the first day of spring and the first day of summer?

Check your answer before going on to the next topic.

1.7 CALCULATING SPEED

$$\text{Distance} \div \text{Time} = \text{Speed}$$

Speed is calculated as the distance covered in a given amount of time. In 1960, Armin Hary of West Germany ran the 100-meter dash in 10 seconds. His speed was 10 meters per second (100 meters ÷ 10 seconds). Since then, many runners have been faster. In 2008, Usain Bolt of Jamaica ran 10.29 meters per second setting a new world record (100 meters ÷ 9.72 seconds).

You are probably most familiar with speed in terms of how fast a car is traveling. Cars in the United States have a speedometer that shows how many miles the car will travel in 1 hour. A car moving at 40 miles per hour would require 1 hour to complete a 40-mile trip and 2 hours to complete a trip of 80 miles, assuming that the driver did not slow down or increase speed.

Scientists use the metric system of measurement to record speed. The car traveling at 40 miles per hour would cover 64.37 kilometers in 1 hour. Traveling at 90 kilometers per hour sounds fast, but it is equivalent to 55 miles per hour. Common units of measure for speed in metric are meters per second (m/s) and kilometers per hour (km/hr).

CHECK YOUR UNDERSTANDING 1.7 (For answer, see page 150.)

A runner finished a 100-meter race in 12 seconds. What was the runner's speed?

Check your answer before going on to the next topic.

1.8 MEASURING TEMPERATURE

Hot and cold are terms we use to describe temperature compared to our own body temperature. Warm to one person may be hot to another. Scientists need a tool for accurate measurements of heat energy.

Thermometers are used to measure temperature. Thermometers may be a thin tube of glass filled with a liquid or an electronic digital device. Your school nurse may have used a digital thermometer to check your temperature.

In the English system, temperature is measured using the Fahrenheit scale (°F), named for its inventor German physicist Daniel Fahrenheit. This scale sets the freezing point of water at 32°F and the boiling point at 212°F.

Swedish astronomer Anders Celsius devised a scale based on 100 units. The Celsius or centigrade scale sets the freezing point of water at 0°C and the boiling point of water at 100°C. Scientists prefer this system because like the metric system it is based on units divisible by ten.

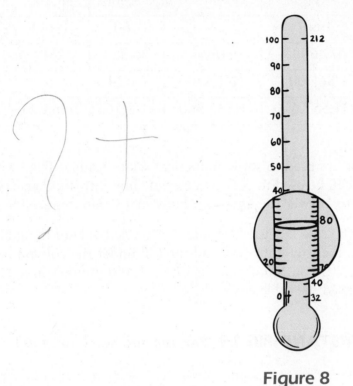

Figure 8

CHECK YOUR UNDERSTANDING 1.8 (For answer, see page 150.)

THE BIG IDEA
Data can be displayed using tables, charts, and graphs.   After making observations or conducting experiments, scientists need to communicate their findings. Tables, charts, and graphs are ways that scientists display numerical data so that others may easily understand the outcome of the experiment or the meaning of the observations.

Look at the thermometers shown in Figure 8. What is the temperature in Fahrenheit and in Celsius?

Check your answer before going on to the next topic.

1.9 TABLES

EXAMPLE OF A TABLE

Units of Time					
Unit	Second	Minute	Hour	Day	Year
Second	1	60	3,600	86,400	31,557,600
Minute	1/60	1	60	1,440	525,960
Hour	1/3,600	1/60	1	24	8,766
Day	1/86,400	1/1,440	1/24	1	365.25
Year	1/31,557,600	1/535,960	1/8,766	1/365.25	1

Tables are used to show numerical relationships. The Units of Time table can be used to convert one time unit into another. Try using the table to answer the question: How many minutes are there in a day?

Find the row labeled "Minute" on the left side of the table. Move across the row until you find the cell under the column heading "Day." If you found that there are 1,440 minutes in a day, you have used the table correctly.

CHECK YOUR UNDERSTANDING 1.9 (For answer, see page 150.)

How many minutes are in a year?

_____

Check your answer before going on to the next topic.

1.10 LINE GRAPHS

Line graphs are used when an event continues uninterrupted over a period of time. It may be difficult or impossible to record all the data in a given period of time. Scientists take samples during the observation or experiment. They plot the data points on a graph and then draw a line that fits the pattern of the dots. Think of it as connecting the dots.

Example of a Line Graph

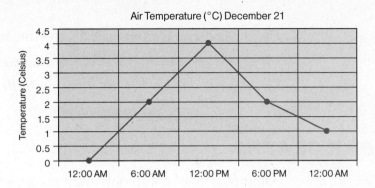

The title of the table indicates that the data was collected December 21. The *y*-axis (left vertical side of the graph) shows the temperatures in degrees Celsius. The *x*-axis (bottom horizontal edge of the graph) shows the time of day each temperature sample was recorded. The graph shows that the temperature depended on the time of day. Line graphs are read as *y* depends on *x*.

The graph clearly shows the temperature trend during the 24-hour period. Because temperature is continuous, we can estimate the air temperature at other times of the day. For example, there are 6 hours between 12 A.M. and 6 A.M. Half way between the two times is 3 A.M. According to the graph the temperature at 3 A.M was 1°C.

The scientist could have recorded the data as a table. It is more difficult to estimate temperatures at other times of the day from the data table.

A technology that scientists often use is an electronic spreadsheet. The numbers are converted into a graph by a special computer program. The spreadsheet used to make the temperature graph is shown below.

Time of Day	Temperature (°C) December 21
12:00 A.M.	0
6:00 A.M.	2
12:00 P.M.	4
6:00 P.M.	2
12:00 A.M.	1

CHECK YOUR UNDERSTANDING 1.10 (See answer, page 150.)

Use the line graph from page 13 to estimate the temperature at 9 A.M.

Check your answer before going on to the next topic.

1.11 BAR GRAPHS

Another type of commonly used graph is the bar graph. Bar graphs compare values from different events. Precipitation is a weather event that can be measured. The example of a bar graph shows the snowfall for December 2008. The graph shows how much snow fell and the dates when the events occurred.

Example of a Bar Graph

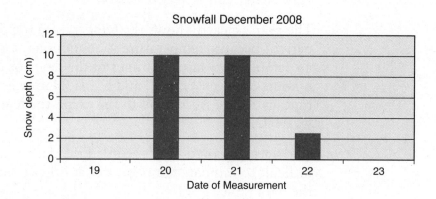

CHECK YOUR UNDERSTANDING 1.11 (For answer, see page 150.)

Why are there no bars for December 19 and December 23?

Check your answer before going on to the next topic.

1.12 CIRCLE GRAPHS

Circle graphs or pie charts are used to show the parts of a whole based on their fraction of the total. Scientists use circle graphs to provide a comparison of smaller sets to a larger set. For example, during a year, precipitation may be in the form of rain, snow, sleet, or freezing rain. The large set is number of days of precipitation in a year. The smaller sets are days of rain, snow, sleet, and freezing rain. The sample circle graph gives a picture of how important each type of precipitation is at a specific location.

Sample Circle Graph

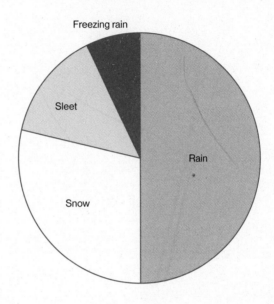

Precipitation Events Per Year

CHECK YOUR UNDERSTANDING 1.12 (For answer, see page 150.)

What type of event accounts for half of the total number of days of precipitation?

Check your answer before going on to the next topic.

EXPERIMENTAL DESIGN

EXPERIMENTAL DESIGN CONCEPT MAP

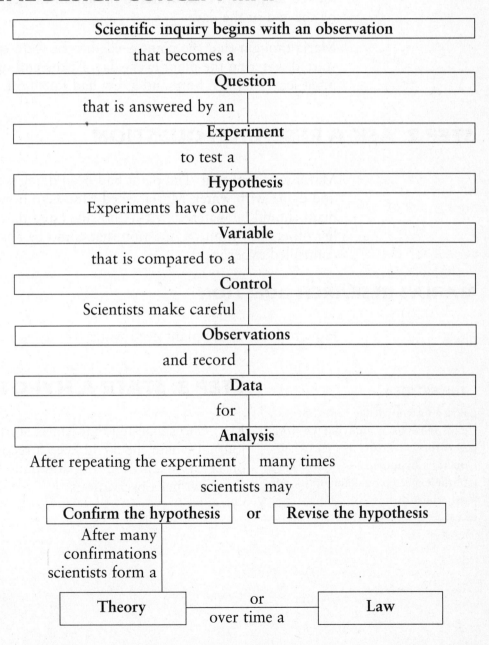

Scientific inquiry begins with an observation

that becomes a

Question

that is answered by an

Experiment

to test a

Hypothesis

Experiments have one

Variable

that is compared to a

Control

Scientists make careful

Observations

and record

Data

for

Analysis

After repeating the experiment   many times

scientists may

Confirm the hypothesis   or   **Revise the hypothesis**

After many
confirmations
scientists form a

Theory   or   **Law**
over time a

STEP 1: MAKE AN OBSERVATION

Mrs. Cho's class has a science corner with a plant. Maria's class job is to water the plant. Maria decided to water the plant once a week starting with Monday.

Monday of the second week, Maria noticed that the plant leaves looked limp. After she watered the plant, the leaves seemed normal by the end of the day.

Maria thought that the plant needed to be watered more often. She started watering the plant every day. At the end of the week, the plant leaves looked limp and a few had fallen off.

STEP 2: ASK A RESEARCH QUESTION

Maria was confused. The plant had been dying without water and dying with water. Maria asked Mrs. Cho how often the plant should be watered. Mrs. Cho suggested that Maria turn her observations into a question that could be answered by a controlled experiment.

MARIA'S RESEARCH QUESTION

How often does the plant need water to stay healthy?

STEP 3: STATE A HYPOTHESIS

Mrs. Cho suggested that Maria start a research journal where she could keep a record of her observations, procedures, and data. In her journal, she drew her observations.

MARIA'S OBSERVATIONS

Plant with limp leaves	Normal plant	Plant with falling leaves
Plant on Monday	Plant after water on Monday	Plant after watering every day

Maria found a plant book at the library. She discovered that her plant needed to be in soil that was moist, but not wet. Maria wrote this information in her journal.

Maria analyzed her observations. When she watered the plant only on Monday, she remembered that the soil was very dry. When she watered the plant everyday, she recalled that the soil had been very wet. She added this information to her observations. Finally Maria wrote her hypothesis.

MARIA'S HYPOTHESIS

If the plant is watered on Monday and Thursday, the soil will stay moist, and the plant will be healthy.

> **THE BIG IDEA**
>
> In designing an experiment, a scientist changes only one condition at a time to find out what is causing the observed event.

STEP 4: SET UP THE CONTROL AND THE VARIABLE

Maria talked to Mrs. Cho about the experiment. Mrs. Cho explained that experiments should be controlled so that the scientist proves that one factor is causing the change.

The **variable** is the condition that is **changed**. The **control** keeps other factors **the same** in order to prove that the variable is causing the change.

Maria decided that she needed three plants. Maria made a table in her journal of conditions for the experiment. She decided what conditions she would keep the same (control) and what condition she would change (variable).

MARIA'S JOURNAL ENTRY

Conditions	Plant #1	Plant #2	Plant #3
Sunlight	Same	Same	Same
Water per week	One time	Two times	Five times
Soil	Same	Same	Same
Air humidity (water in the air)	Same	Same	Same
Warm air temperature	Same	Same	Same

The variable would be how often the plants were watered.

THE BIG IDEA

Scientists keep a record of their procedures so that the experiment can be repeated or changed as needed.

STEP 5: FOLLOW THE PROCEDURE

In her journal, Maria drew a diagram of her experimental plan. Maria planned to start with three plants that needed water.

MARIA'S JOURNAL ENTRY

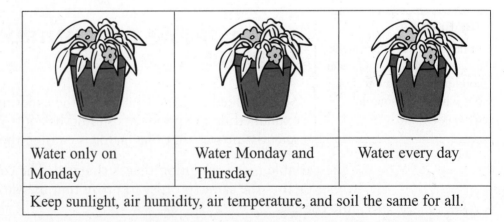

Water only on Monday	Water Monday and Thursday	Water every day
Keep sunlight, air humidity, air temperature, and soil the same for all.		

Maria showed her journal to Mrs. Cho. Mrs. Cho had several questions about the set-up. Would the amount of water be the same each time? How much water would be used? What would Maria use to measure the water?

To answer the questions, Maria needed to know how much water was needed to make the soil moist, but not too wet.

Maria filled a 100-milliliter graduated cylinder with water. She poured water from the cylinder onto the plant soil surface until water drained out onto the tray under the plant pot. From this she estimated that 50 milliliters of water would make the soil moist, but not too wet. Maria revised her procedure to show the amount of water used each time.

To control the other factors, Mrs. Cho helped Maria find a space in the classroom where the three plants would have the same amount of light and air temperature. Mrs. Cho was able to borrow the same type of plant from other teachers. Maria was ready to start the experiment.

> **THE BIG IDEA**
>
> Scientists carefully record data and detailed observations.

STEP 6: COLLECT DATA

Maria started her experiment on a Monday. She kept careful records in her journal.

Week 1

Day	Plant 1	Plant 2	Plant 3
Monday	Soil was dry Leaves were drooping Add 50 ml water	Soil was dry Leaves were drooping Add 50 ml water	Soil was dry Leaves were drooping Add 50 ml water
Tuesday	Soil was moist Leaves were normal No water added	Soil was moist Leaves were normal No water added	Soil was moist Leaves were normal Add 50 ml water
Wednesday	Soil was moist Leaves were normal No water added	Soil was moist Leaves were normal No water added	Soil was wet Leaves were drooping Add 50 ml water
Thursday	Soil was dry Leaves were normal No water added	Soil was dry Leaves were normal Add 50 ml water	Soil was wet Leaves were drooping Add 50 ml water
Friday	Soil was dry Leaves were limp No water added	Soil was moist Leaves were normal No water added	Soil was wet Leaves were rolling up, one leaf fell off Add 50 ml of water

Week 2

Day	Plant 1	Plant 2	Plant 3
Monday	Soil was dry Leaves were drooping Add 50 ml water	Soil was moist Leaves were normal Add 50 ml water	Soil was moist Two leaves fell off Add 50 ml water
Tuesday	Soil was moist Leaves were normal No water added	Soil was moist Leaves were normal No water added	Soil was wet Another leaf fell off Add 50 ml water
Wednesday	Soil was moist Leaves were normal No water added	Soil was moist Leaves were normal No water added	Soil was wet Leaves were rolled up Add 50 ml water
Thursday	Soil was dry Leaves were normal No water added	Soil was dry Leaves were normal, new leaves were starting to grow Add 50 ml water	Soil was wet Another leaf fell off Add 50 ml water
Friday	Soil was dry Leaves were limp No water added	Soil was moist Leaves were normal No water added	Soil was wet Leaves were rolling up Add 50 ml water

After two weeks, Maria ended her experiment. She tried watering all three plants using the Monday and Thursday schedule. Plant 1 started to grow new leaves. Plant 2 was noticeably taller than plant 1 and 3. Plant 3 returned to normal.

THE BIG IDEA
Scientists use their results to confirm the hypothesis or to revise the hypothesis.

STEP 7: ACCEPT OR REVISE THE HYPOTHESIS

CONFIRMING THE HYPOTHESIS

If the results of the experiment match the hypothesis, a scientist will conduct more experiments. If the results are the same, a scientist will conclude that the hypothesis was correct. The next step is to share the experimental design and results with other scientists who will run more tests.

REVISING THE HYPOTHESIS

If the results are unexpected, a scientist will conduct more experiments. If the same unexpected results occur, the scientist will consider other possible causes and form a new hypothesis. This will start a new series of investigations. This process may also occur if other scientists cannot obtain the same results.

Maria's hypothesis was that watering the plant on Monday and Thursday would keep the soil moist and the plant healthy. After analyzing her data, she found that the soil dried out by Thursday, but the plant was healthier than watering once a week or watering every day. Her experimental results did not completely support her hypothesis. She needed to revise her hypothesis but was not sure what change to make.

> ### THE BIG IDEA
> Scientific theories are suggested explanations and are subject to testing and revision. When a theory becomes accepted as an explanation for every example of a specific natural event it becomes a scientific law.

Mrs. Cho suggested that Maria ask her classmates for ideas. She presented her experiment to the class. Sasha suggested that maybe the soil was moist down by the roots even though the surface was dry. She said her mom had a special tool to test plant soil for water. Sasha volunteered to work with Maria to try the experiment again this time using a tool to measure the soil moisture.

THEORY

If the results of experiments continue to be the same, a scientist may form a **theory**, an explanation of why something happens. A theory may become accepted if other researchers prove the hypothesis through experiments. Other scientists may contribute ideas and proofs to expand a theory. Theories must be testable and supported by experience-based observations.

LAW

Over many years of testing by different scientists, a discovery may become a **law**, a statement that applies to all examples of specific events.

Maria's experiments were limited to a specific type of plant under the conditions of her classroom. Her watering solution might not work under other conditions of sunlight, air temperature, and room humidity (water in the air). Therefore, her work would end with the experimentation phase to find the best watering schedule for her situation. She would not form a theory or develop a law from her work.

BIG IDEA CHECK-UP (For answers, see page 150.)

Mr. Kane's class wanted to know how much water a certain type of fast-growing plant needed to develop from seed to healthy plant. The students were divided into three groups. Each group formed a hypothesis about the amount of water needed for best results. Their experimental setup is shown in the illustration, and the results are in the table.

| Group 1 | Group 2 | Group 3 | Group 4 |

Group	Amount of Water	Experimental Results Week 1	Week 2
1	2 ml of water every day	5 out of 10 seeds sprouted. Plants were yellow-white and had two leaves.	All plants died.
2	2 ml of water every other day	9 out of 10 seeds sprouted. Plants were green and had two leaves each.	Plant stems were 10 to 15 mm in length. Plants had 4 large leaves. No plants died.
3	2 ml of water once each week	7 out 10 seeds sprouted. Plants were green and had two leaves.	Plants stems were 5 to 10 mm in length. Plants had 4 small leaves. No plants died.
4	No water	0 out of 10 seeds sprouted.	0 out of 10 seeds sprouted.

Use the information in the illustration and table to answer the questions.

Multiple Choice (Circle the correct answer.)

1. What tool was used to measure the amount of water?

    A. Ruler

    B. Thermometer

    C. Graduated cylinder

    D. Calendar

2. If the amount of water is the variable, what is the control for the experiment?

    A. Water every day

    B. Water every other day

    C. Water once a week

    D. No water

Open-Ended Question

3. Write a conclusion based on the experimental results.

    _____

    _____

    _____

    _____

    _____

    _____

    _____

Check your answers before going on to the next topic.

LIFE SCIENCE

LIFE SCIENCE CONCEPT MAP

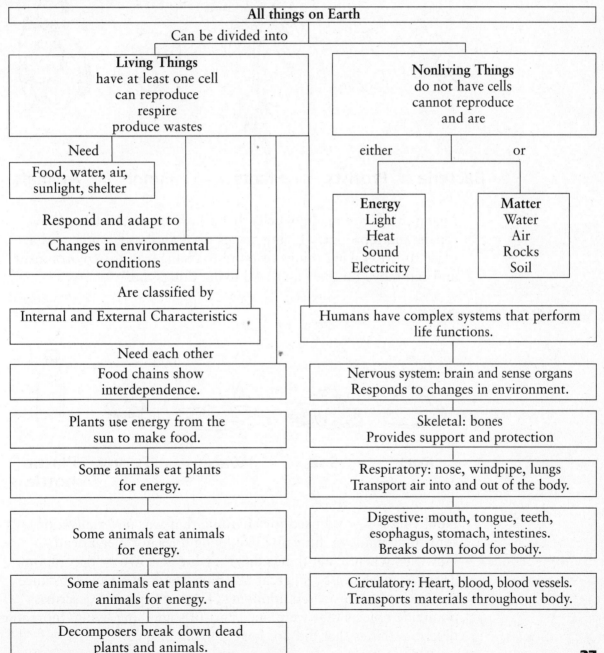

All things on Earth

Can be divided into

Living Things
have at least one cell
can reproduce
respire
produce wastes

Nonliving Things
do not have cells
cannot reproduce
and are

Need

Food, water, air,
sunlight, shelter

Respond and adapt to

Changes in environmental
conditions

Are classified by

Internal and External Characteristics

Need each other

Food chains show
interdependence.

Plants use energy from the
sun to make food.

Some animals eat plants
for energy.

Some animals eat animals
for energy.

Some animals eat plants and
animals for energy.

Decomposers break down dead
plants and animals.

either   or

Energy
Light
Heat
Sound
Electricity

Matter
Water
Air
Rocks
Soil

Humans have complex systems that perform
life functions.

Nervous system: brain and sense organs
Responds to changes in environment.

Skeletal: bones
Provides support and protection

Respiratory: nose, windpipe, lungs
Transport air into and out of the body.

Digestive: mouth, tongue, teeth,
esophagus, stomach, intestines.
Breaks down food for body.

Circulatory: Heart, blood, blood vessels.
Transports materials throughout body.

3.1 LIVING VS. NONLIVING

THE BIG IDEA
All things on Earth are either living or nonliving.

LIVING THINGS

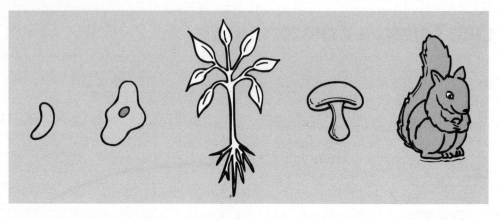

Bacteria   Protists   Plants   Fungi   Animals

Living things are made of cells that allow the organism to grow, repair, and reproduce. Living things need air, food, water, and a place to live. Living things respond to changes in the environment and can move on their own. All living things die.

NONLIVING THINGS

Rock   Soil   Water   Penny   Plastic bottle

Nonliving is a word used for anything that was never alive, or something made by humans. Nonliving does not mean dead. However, when living things die, they break down or decompose. These changes can be so great that they result in a new substance that might be considered nonliving. The term **organic** describes nonliving objects that were once part of something living. **Inorganic** describes things that did not have a living origin.

BIG IDEA CHECK-UP 3.1 (For answers, see page 151.)

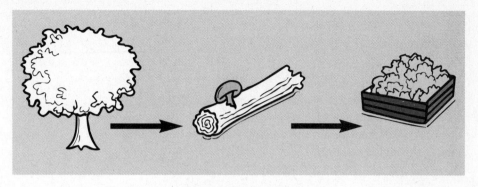

Tree   Log with mushroom   Compost
               and bacteria

Multiple Choice (Circle the correct answer.)

1. Which of the following is an example of something that is nonliving?

    A. Log

    B. Tree

    C. Compost

    D. Mushroom

2. Which statement would you use to decide that a sea sponge is living thing?

    A. It soaks up water.

    B. It reproduces to make new sea sponges.

    C. It releases air bubbles when it is squeezed.

    D. It turns brown when it dries out.

Check your answers before going to the next topic.

3.2 ENERGY FLOWS AND FOOD CHAINS

> **THE BIG IDEA**
>
> Energy flows through a system. Plants obtain their energy directly from the Sun. Some animals obtain energy by eating plants, while others obtain energy by eating other animals. This is a food chain. **A food chain shows the flow of energy through a system.**

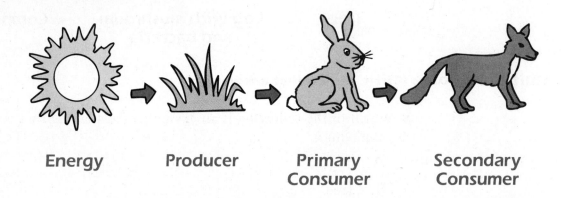

Energy   **Producer**   **Primary Consumer**   **Secondary Consumer**

The diagram shows the flow of energy through a system. The **Sun** is the primary source of energy for all living things on Earth. That is why the arrow begins with the Sun.

Plants are able to use energy from the Sun to make food for themselves and others. They are called **producers**. To produce means to make something.

Some animals only eat plants. They are called **herbivores** *(er-bi-vors)* from the Latin word for "plant eater." Because herbivores obtain energy directly from producers they are called **primary consumers** (first users).

Other animals eat animals for food. They are called **carnivores** *(car-ni-vors)* from the French word for "meat eater." Carnivores are the second users of energy from producers; therefore, they are called **secondary consumers** or, if nothing eats them, a **top consumer**.

BIG IDEA CHECK-UP 3.2 (For answers, see page 151.)

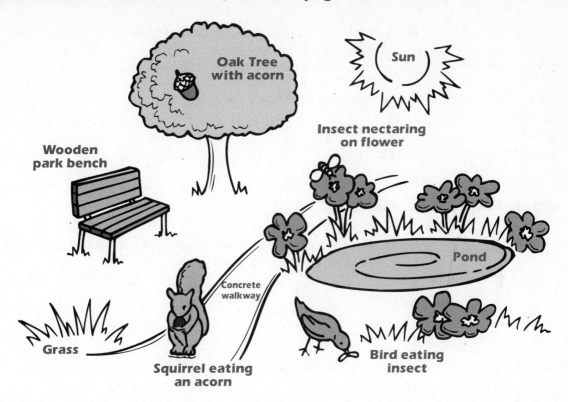

Multiple Choice (Circle the correct answer.)

1. What is the source of energy for the system shown in the picture of the park?

   A. The oak tree

   B. The water

   C. The grass

   D. The Sun

2. Which pair of objects is an example that has both a living object and a nonliving object?

   A. The squirrel and acorn

   B. The oak tree and wooden park bench

   C. The bird and insect

   D. The insect and flower

Sun → flower → insect → bird

3. A producer makes food for itself and others. In the food chain above, what is the producer?

   A. Sun

   B. Flower

   C. Insect

   D. Bird

Open-Ended Questions

4. Snakes that eat birds escape from a pet store into the wild. Draw the new food chain in the box below.

5. What will happen to the insect population if the snakes become part of the community? Write your answer on the lines below.

   _____

   _____

   _____

   _____

   _____

   _____

Check your answers before going on to the next topic.

3.3 PLANTS VS. ANIMALS

WHAT ARE THE DIFFERENCES BETWEEN PLANTS AND ANIMALS?

Plants and animals are both living things. Like all living things they both need air, water, and nutrients. However, plants and animals use different parts of the air, process water differently, and obtain nutrients differently.

Plants make food for themselves and others as producers in a food chain. Animals obtain their energy by eating plants or animals that eat plants. Do you eat both vegetables and meat? You are an **omnivore**.

Plants use energy from the Sun to make food, but they also need other nutrients from the soil. Plants have special structures for transporting nutrients. Animals obtain nutrients by eating and digesting plants or other animals.

Venn Diagram of Plants vs. Animals

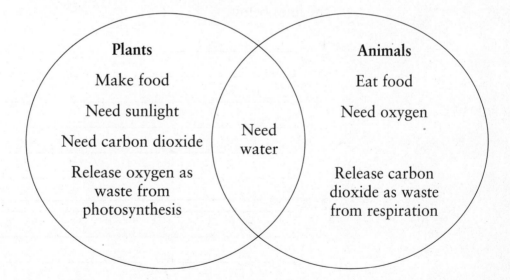

Plants
Make food
Need sunlight
Need carbon dioxide
Release oxygen as waste from photosynthesis

Need water

Animals
Eat food
Need oxygen
Release carbon dioxide as waste from respiration

BIG IDEA CHECK-UP 3.3 (For answers, see page 152.)

Multiple Choice (Circle the correct answer.)

1. Which of the following do both plants and animals need to live?
   A. carbon dioxide
   B. oxygen
   C. soil
   D. water

2. Which gas do animals produce as waste from respiration?
   A. oxygen
   B. carbon dioxide
   C. water
   D. nitrogen

Open-Ended Question

3. List two reasons that animals need plants. Write your answer on the lines below.

   1. _____
      _____
      _____
      _____
      _____
      _____

   2. _____
      _____
      _____
      _____
      _____
      _____

Check your answers before going on to the next topic.

3.4 STRUCTURES OF PLANTS

THE BIG IDEA
Most plants have special structures to make food, obtain water, exchange gases, and reproduce.

WHAT ARE THE CHARACTERISTICS OF PLANTS?

Most, but not all, plants have leaves, stems, roots, and seeds.

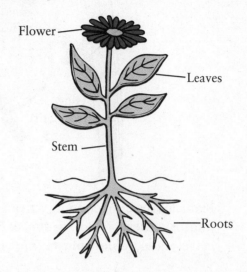

Leaves are important for photosynthesis. They take in carbon dioxide and release oxygen.

Stems move water and nutrients through the plant.

Seeds contain tiny plants that sprout under the right conditions.

Roots obtain water and nutrients.

Flower Anatomy

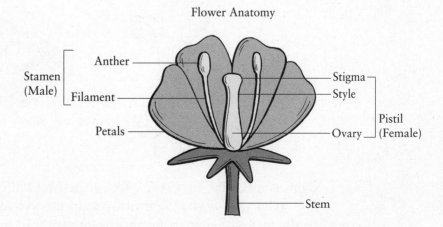

Some plants have special structures that produce seeds.

HOW DO THE PARTS OF PLANTS WORK TOGETHER?

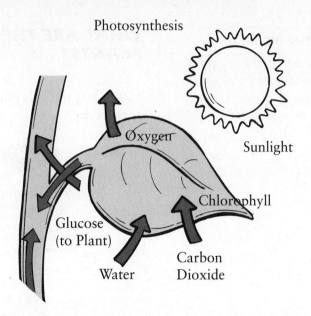

Plant **leaves** collect energy from the Sun to make food by **photosynthesis** *(foto-sin-the-sis)*. Special structures inside the leaves called **chloroplasts** *(klor-o-plasts)* respond to sunlight.

Openings on the leaves called **stoma take in carbon dioxide** from the air for photosynthesis. **Stoma release oxygen** into the air. Oxygen is a waste product of photosynthesis.

Water can also move out of a leaf opening. This helps pull water up the stem. Water helps a plant with photosynthesis and helps a plant keep its shape.

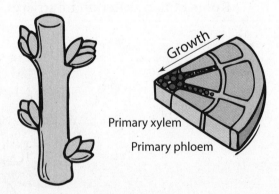

Plant stems have tiny straw-like structures called **phloem** *(flow-um)* and **xylem** *(zy-lum)*. Water moves up the xylem from the roots to the leaves. Sugar from photosynthesis and nutrients move through the phloem. Stems help support the plant.

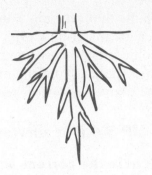

Roots absorb water and nutrients for the plant. Some plants have roots in the soil. Other plants have roots that take water out of the air. The root does not have structures for photosynthesis. Pressure in the root can force water up the stem.

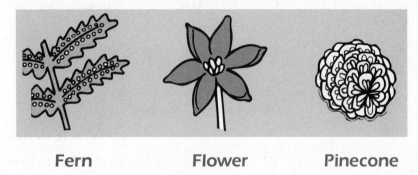

Fern　　　　　**Flower**　　　　　**Pinecone**

Many plants reproduce by making seeds. Seeds contain tiny plants that can grow when conditions are right. Some plants make seeds in the flowers. Others, such as pinecones, have naked seeds because they do not come from a flower and are not covered. Mosses and ferns make spores that are not seeds because there is no plant inside, only the instructions for making a new plant.

Life Cycle of a Flowering Plant

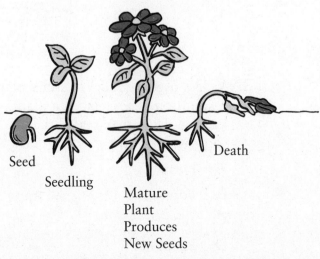

Seed

Seedling

Mature
Plant
Produces
New Seeds

Death

Plants that reproduce by spores depend on water to carry their genetic material. Conifers such as pine trees rely on the wind for pollination. Flowering plants attract insects, such as honeybees, that carry pollen from the male part of the plant to the female part of the plant. The insect receives sweet nectar from the flower.

BIG IDEA CHECK-UP 3.4 (For answers, see page 152.)

Multiple Choice (Circle the correct answer.)

1. What part of the plant uses sunlight to make food by photosynthesis?

   A. The roots

   B. The leaves

   C. The stem

   D. The seeds

2. A plant was placed under a box for two days. What would be most likely to happen?

   A. The plant would not be able to make food.

   B. The plant would not be able to produce flowers.

   C. The plant would not be able to store water.

   D. The plant would not be able to obtain carbon dioxide.

3. A class used a model to study plant photosynthesis and animal respiration. They placed a plant, soil, water, and an earthworm in a jar. Then they sealed the jar. What else did the students need to provide for the model to be a complete system?

   A. Heat

   B. Food

   C. Air

   D. Sunlight

Open-Ended Question

4. An apple orchard depends on one honeybee colony per acre for pollination. What would happen if there were fewer honeybees?

_____

_____

_____

_____

Check your answers before going on to the next topic.

3.5 STRUCTURES OF ANIMALS

What do animals need?

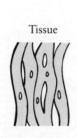

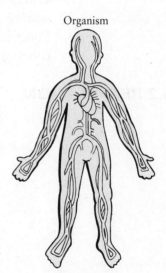

Organism

Organ

Tissue

Cell

Heart
Muscle
Cell

Heart Muscle
Tissue

Heart

Circulatory System

THE BIG IDEA

Most animals have special structures for digesting food, exchanging gases, excreting waste, moving, sensing environmental conditions, and transporting nutrients.

An organism is a living thing that can be a single cell or made of many cells. **Animals are multicellular (many cells).** Because animals have more than one cell, they need systems to be able to do their life processes. When a group of cells works together for the same job, tissue is formed. When different types of tissues work together, they form an organ. Organs can work together to form an organ system. Not all animals have organs and organ systems. But they all have tissues.

3.5.1 DIGESTION

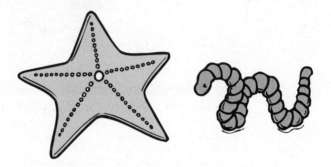

Whether they are herbivores or carnivores, animals must obtain nutrients by digesting food. Animals such as the starfish only have one opening in their digestive systems. A starfish uses its arms to open a clam shell. Then it inserts a tube into the shell and vacuums up the clam. After nutrients from the clam are digested, the waste passes out the same opening that food entered.

Other animals such as the worm have two openings and a digestive system that is like a long tube. Worms take nutrients out of soil. Food enters the mouth of the worm and passes along the digestive tube, and waste is excreted at the other end.

3.5.2 RESPIRATION

To burn food, animals need oxygen. Some animals such as the worm breathe through their skins. Others, such as frogs, have tiny lungs but also use their skin for breathing. Fish have special structures called gills that remove oxygen and get rid of carbon dioxide. Bats have lungs. Lungs and gills are part of a respiratory system.

3.5.3 MOVEMENT AND PROTECTION

Because animals have more than one cell, they need a system that provides support and protection. Some soft-bodied animals have protective structures such as spines or shells. The shell of the octopus is under its skin. The shell of the clam is on the outside.

A skeleton provides support and helps with movement. Some animals have skeletons that are on the outside, for example, the grasshopper, crab, and spider. Other animals such as the bat, snake, and frog have bones that form a skeleton on the inside. Muscles work with the skeleton so that animals can move.

3.5.4 CIRCULATION

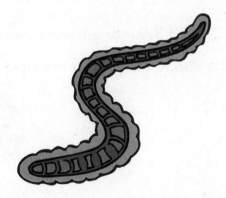

Animals also need a system for transporting nutrients, water, and gases around the body and for getting rid of waste. Some animals have hearts that pump blood through a system of blood vessels. The earthworm has a simple circulatory system.

3.5.5 RESPONSE

Most animals move at some time during their lives and have structures for movement, protection, and collection of information about the environment. Eyes, ears, skin, nerves, noses, and taste buds are ways that animals can gather information about the environment. Fish, for example, have a strong sense of smell. Owls have excellent eyesight. Bats can avoid flying into objects because of their excellent hearing.

3.5.6 REPRODUCTION

Animals pass on traits by reproduction. The characteristics of the parents are passed on to the offspring. Half the traits are inherited from each parent. Only those things that are coded in deoxyribonucleic acid (DNA) on genes can be inherited.

BIG IDEA CHECK-UP 3.5 (For answers, see page 153.)

Multiple Choice (Circle the correct answer.)

1. Snails have a soft body. What is the job of the shell?
   A. The shell helps the snail breathe.
   B. The shell helps the snail digest food.
   C. The shell helps the snail move.
   D. The shell helps protect the snail.

2. After a heavy rainstorm, earthworms crawl onto the sidewalk. Which statement best explains the response to the wet soil?

   A. Earthworms breathe through their skin.

   B. Earthworms digest food from soil.

   C. Earthworms have a simple circulatory system.

   D. Earthworms do not have a skeleton.

Check your answers before going on to the next topic.

3.6 HUMAN BODY SYSTEMS

THE BIG IDEA

The human body is organized into systems that work together for survival.

Humans are animals because they have many cells that are organized into tissues, must obtain food for survival, and are able to move around in response to change in their surroundings. Human bodies are very complex. Tissues, such as nerve, muscle, and skin, work together in organs. Organs work together in systems that include the nervous, skeletal, respiratory, digestive, and circulatory systems. There are more human systems, but these systems help humans meet basic survival needs.

3.6.1 NERVOUS SYSTEM

The nervous system includes nerves, sensory organs, and the brain. Nerve cells collect information from the environment and send signals to the brain. The brain makes sense of the information and sends back commands through the nerves to other organ systems so that the body can react. Your brain has many specialized parts that receive, process, store, and send information.

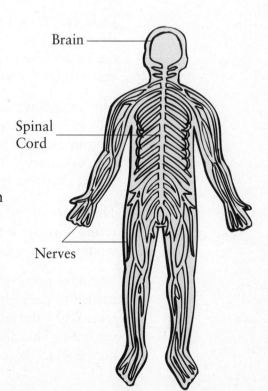

Brain

Spinal Cord

Nerves

3.6.1.A The Brain

The central nervous system is composed of the brain and spinal nerves. The brain controls all of the body's functions. One part of the brain controls your breathing, digestion, and heartbeat. You don't even think about inhaling and exhaling, breaking down food, or contracting and relaxing of your heart muscle. Your brain automatically controls respiration, digestion, and circulation whether you are awake or sleeping. This part of the nervous system is called the **autonomic nervous system**. You could think of it as automatic.

Other parts of the brain process information that you receive from sensory organs and make decisions on how you will react. The parts of the brain that allow you to think, read, write, speak, and move are the **somatic nervous system**. Somatic means body.

3.6.1.B Sensory Organs

Sight

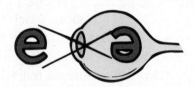

Light reflects off the letters of this page. The light enters your eyes through a lens. Notice in the diagram of the eye that the image on the retina (the back of the eye) is upside down. This happens because your lens is curved.

The image on the retina is converted to electrical signals by special cells called rods and cones. The signal travels along the optic nerve to your brain. The brain decodes and interprets the message from the optic nerve. You are then able to understand the meaning of the letters that you read.

The cones of the retina allow you to see color. The rods of the retina allow you to see motion. Movement is seen in black and white. You can only see objects when they reflect light.

You have two eyes that work together. The image from the left eye is carried to the right side of the brain. The image from the right eye is received by the left side of the brain. Both sets of images are merged in the brain. This allows you to see objects in three dimensions and decide which objects are near and which are far from you.

Hearing

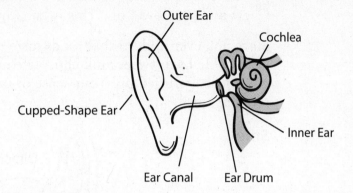

The slightly cupped shape of the ear collects sound waves. The sound waves are focused through the narrow channel of the auditory canal. They strike the eardrum causing it to vibrate. Three tiny bones, the smallest in the human body, carry the sound to the snail-shaped cochlea. Fiber-like hairs of the inner ear translate the sound signals to electrical signals for nerves that connect to the brain, where the information is processed for meaning and action.

You are sitting in a quiet room reading. A fire alarm sounds. What happens? Your ears collect the sound waves. Your brain recognizes the sound of the alarm and remembers that it means danger. You stand and walk out of the room.

Standing and walking are things that you think about and control. If you were startled by the sound, your heart rate may have increased. This is something that you do not control. It happens as a response to possible danger.

Taste

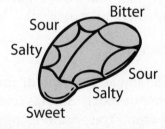

Lemons are sour, sugar is sweet, salt is salty, and lemon peels are bitter. Sour, sweet, salty, and bitter are basic categories of taste recognition. The taste buds of the tongue gather most of the information about the taste of a substance. Cells of the palate (roof of the mouth) and the back of the throat also play a small role in sensing taste. The sensory cells send messages to the brain along nerves. The brain then sends back messages such as: "This is good. Chew and swallow." or "This is bad. Spit it out."

Have you ever noticed that foods taste different when you have a head cold? Even sweet milk chocolate may lose its appeal when your nose is stuffed up. Your sense of taste is enhanced by your sense of smell.

Smell

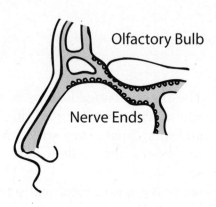

Cookies baking in the oven release chemical molecules into the air. You inhale the air. The molecules enter your nose and stimulate the olfactory bulb, the organ of smell. Electrical signals travel to the brain along the olfactory nerve. Your brain recognizes the smell of cookies and remembers that cookies taste good. Perhaps your mouth starts to water. Maybe you respond by walking to the kitchen in hopes of enjoying a freshly baked treat.

Throughout your life, you build memories based on aromas. The memory of sour milk protects you from drinking something harmful. The memory of a ripe banana tells you that the food is safe to eat. Some odors alert you to danger; for example, smoke means fire, and a strong, foul odor could be a skunk. Odor memories are important for safety and survival.

Touch

Walking across sand at the shore during the summer stimulates your sense of touch. Special cells in your skin feel pressure from the grains of sand. Your brain recognizes that the surface is soft and will make adjustments to help you keep your balance as the grains of sand move out from under your feet.

Other cells in your skin will send a message to your brain about the temperature of the sand. Your brain will let you know that you need to walk faster to avoid pain from the hot sand.

Different skin cells collect information about texture. Your brain recognizes that the sand is rough and abrasive.

Pressure, temperature, and texture are all examples of environmental cues to which humans respond. After determining that the sand is hard to walk on, hot, and rough, your brain may help you make the decision to wear beach shoes instead of going barefoot.

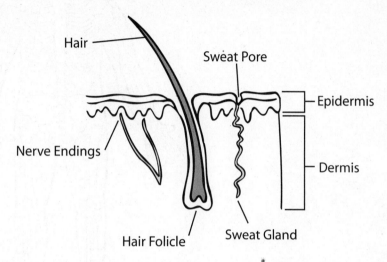

Skin is the largest organ of the human body. The types and amount of sensory cells in the skin depend on the location of the skin. There are more pressure and pain cells in the skin of your hands than in the skin on the back of your neck.

Have you ever noticed that your hairs stand up when you have goose bumps? Hairs on your skin are connected to nerves. When an insect lands on your arm, you feel it mostly because the insect's body moves the hairs of your skin. Your brain may send back a signal to swat the insect or brush it away.

3.6.2   SKELETAL SYSTEM

Humans have an internal skeleton that supports and protects the body from the inside. The skeleton is made of bones that are covered by layers of skin, fat, and muscles.

Look carefully at the human skeleton on page 48. It looks the same on the left side as it does on the right side. This is called **bilateral symmetry**, which means that both halves of the body have the same shape. The skeleton gives the body shape.

Bones are the main organs of the skeleton. The longest bone of the human body is the femur, the bone of the upper leg. The smallest bones are inside the ear. There are four main types of bones: flat bones, long bones, short bones, and irregular bones.

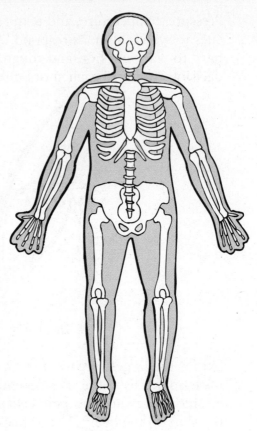

Flat bones can bend just like a sheet of paper to form container-like spaces. Ribs are flat bones that protect the lungs and heart. The flat bones of the skull protect the brain.

Long bones of the arms and legs produce blood cells, store minerals, allow for movement, and provide support for the body. These types of bones are strong like a sheet of paper rolled into a tight tube.

Short bones provide flexibility and fine movement. Short bones are found in the wrists and hands as well as the ankles and feet.

Irregular bones such as those of the backbone protect the spinal nerves and allow twisting motion. Another set of irregular bones is found in the hip. Their shape allows you to stand upright.

Tendons attach the muscles to the bones. Muscles move the bones like puppet strings move a marionette. The brain sends signals through nerves that cause the muscles to contract and relax. The muscles allow you to bend at joints or places where bones come together, for example, your elbows and knees.

3.6.3 RESPIRATORY SYSTEM

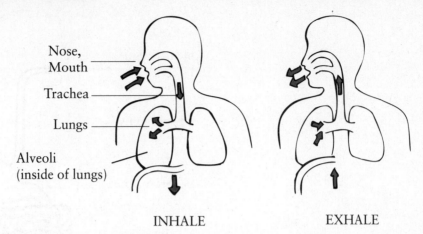

Nose, Mouth
Trachea
Lungs
Alveoli (inside of lungs)

INHALE   EXHALE

You need to obtain oxygen from air and to get rid of carbon dioxide produced by your cells. The main structures of the respiratory system are the nose, **trachea** *(tray-key-a)*, and lungs.

When the **diaphragm** *(die-a-fram)*, a thin muscle under the rib cage, contracts, it causes the space inside the chest to became larger. Air rushes into your nose, flows down your trachea (windpipe), and inflates your lungs.

Air is warmed and moistened as it enters your nose. Tiny hairs inside the nose filter out large particles. Mucus traps dust and pollen.

The cleaned and warmed air passes into the pharynx or back of the throat. There a flap of skin called the epiglottis opens up the trachea. The windpipe splits into two branches called bronchial tubes. One tube goes to each lung.

The bronchial tubes split into smaller and smaller branches in each lung. They end in small clusters of special cells called **alveoli** *(al-vee-o-lee)*. The alveoli look like a bunch of grapes, but each individual alveolus is a small sac that allows oxygen to pass into tiny blood vessels called **capillaries** *(cap-ill-air-ees)*. As oxygen leaves the lungs, carbon dioxide crosses from the capillaries into the lungs.

The diaphragm relaxes. The inside of the chest becomes smaller. Air that now contains carbon dioxide rushes out through the bronchial tubes, flows through the trachea, and exits through the nose. This is exhaling.

Inhaling and exhaling are the two steps of breathing. Your brain takes care of this for you.

3.6.4 DIGESTIVE SYSTEM

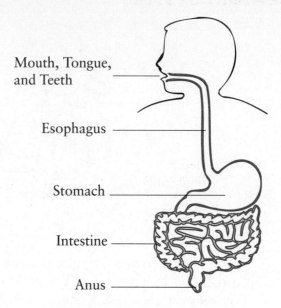

Mouth, Tongue, and Teeth

Esophagus

Stomach

Intestine

Anus

The energy that your body needs comes from the digestion of food. The main structures of the digestive system are the mouth, teeth, tongue, esophagus *(e-sofa-gus)*, stomach, small intestine, and large intestine.

Food is ground up by your teeth as you chew. Chemicals that break down starch into sugar are released by salivary glands under your tongue. Your tongue mixes the chewed food with saliva before it pushes the food to the back of your throat.

The tiny flap of skin (epiglottis) that opened for air to enter your lungs now flips over to cover the trachea and open up your esophagus. No digestion takes place in the esophagus. Food is moved in one direction by rhythmic contractions of the esophagus. You do not feel these contractions and do not need to think about them. Your brain is in control of this action.

The esophagus connects to your stomach. After the food enters the stomach, it is ground up into even smaller particles by contractions of the stomach muscles. Acids and other chemicals break down proteins. The mixture of food particles and chemicals is called chyme *(kime)*.

Chyme flows into the small intestine. Here more chemicals digest fats. The digested food is moved along by slow contractions of the intestinal muscles. Cells inside the intestine let nutrients pass into the capillaries that surround the intestines. Water and waste are processed by the large intestine before the waste is passed out of the body.

3.6.5 CIRCULATORY SYSTEM

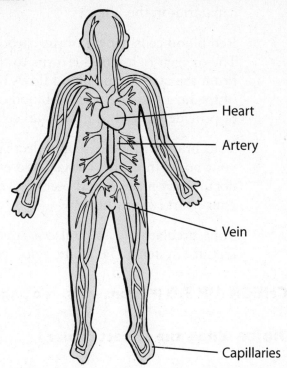

Heart

Artery

Vein

Capillaries

The circulatory system moves substances through the body. The main parts of the circulatory system are the heart, arteries, veins, capillaries, and blood.

The heart is a pump that contracts and relaxes continuously without taking a break. The arteries carry blood away from the heart. Veins carry blood to the heart. Capillaries are tiny connections between the arteries and veins. Red blood cells collect carbon dioxide from the body cells. They exchange the carbon dioxide for oxygen in the lungs.

The brain controls the rate at which your heart beats. When you are resting, you do not need as much oxygen nor do you produce as much carbon dioxide as when you are moving. Your heart rate is slower than when you are moving.

On the playground you use your muscles to run, climb, and jump. The muscles need energy. Blood carries sugars from digestion to your muscles. When your muscle cells burn the sugar for energy, they release carbon dioxide. Red blood cells in capillaries collect the carbon dioxide molecules.

Veins collect blood from capillaries and return it to the heart. Blood enters the upper right chamber of the heart, the right atrium. The heart pumps the blood into the lower right chamber, the right ventricle. The blood leaves the heart through an artery and moves to capillaries in the lungs.

Red blood cells get rid of the carbon dioxide and collect oxygen. The oxygen-rich blood returns to the heart through the left atrium. From the atrium, the blood flows into the left ventricle. A squeeze of the heart sends the fresh blood out to the body through arteries to provide oxygen to individual cells.

Your brain will increase and decrease your heart rate in response to your body's need for oxygen and energy. During a check-up, your doctor may place a stethoscope over your heart to make sure that your heart sounds healthy.

Your circulatory system also carries special cells produced by the immune system. The white blood cells fight infections.

BIG IDEA CHECK-UP 3.6 (For answers, see page 153.)

Multiple Choice (Circle the correct choice.)

1. Which system controls the functions of the body?

   A. Digestive system

   B. Respiratory system

   C. Nervous system

   D. Skeletal system

2. When you eat food, which organ system works closely with your sense of taste?

   A. Sight

   B. Touch

   C. Hearing

   D. Smell

3. Which of the following is not a function of the skeletal system?

   A. Protection of organs

   B. Support for the body

   C. Production of blood cells

   D. Control of breathing

Open-Ended Question

4. Fourth grade students used a monitor that recorded the heart rate and breaths per minute before running, during running, and after running. The average data are shown in the graph below.

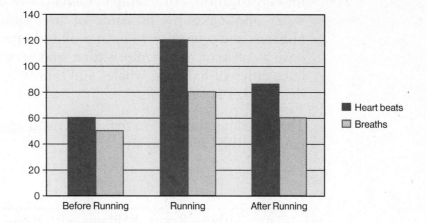

Why did both the heart rate and breathing rate increase during running and decrease after running?

_____

_____

_____

Check your answers before going on to the next topic.

3.7 CLASSIFICATION SYSTEMS

Scientists create classification systems as a way of organizing living things into groups that share similar structures. To sort organisms, scientists make a list of questions that can be answered with a yes or no.

"Does the organism produce food for itself and others by photosynthesis?" This question sorts living things into two groups: things that make food by photosynthesis and things that do not. The "yes" group would be plants.

To sort out the "no" group, another question is needed. "Does the organism have many cells that work together and move to ingest food?" "Yes" to this question would sort out the animals into a group.

The largest grouping of organisms, for example, plant or animal, is called a kingdom. By asking more and more questions, the kingdom can be sorted into smaller and smaller groups. Organisms that are most closely related belong to the same species.

Scientists use common structures instead of where an animal lives for classification. For example, humans live on land, whales live in water, and snakes live on land. All three animals have backbones. Humans and whales have hair; snakes do not. Humans and whales have a constant body temperature; snakes do not. Even though they have different habitats, whales and humans are more closely related than snakes and humans.

The diagrams show ways that plants and animals can be grouped based on their common structures.

PLANTS

Plants without Roots	Plants with Roots
 Rhizome	

ANIMALS

Animals without backbones		Animals with backbones		
Animals without hard coverings	Animals with hard coverings	Animals with feathers	Animals with hair or fur	Animals without feathers, hair, or fur

BIG IDEA CHECK-UP 3.7 (For answers, see page 154.)

Multiple Choice (Circle the correct answer.)

1. The picture shows an animal that was discovered as a fossil in sandstone. What characteristic is shared only with birds?

   A. Wings

   B. Feathers

   C. Scales

   D. Tail

2. Which pair of animals are most closely related?

   A.

   B.

   C.

   D.

Check your answers before going on to the next topic.

3.8 VARIATIONS WITHIN SPECIES

All members of a species share common characteristics, but they do not all look alike. The appearance of an individual depends on the genes inherited from the parents.

Think about humans. You are human. You recognize other humans, but you also recognize individuals based on small differences such as face shape, eye color, hair color, skin color, or height. Even brothers and sisters who have the same parents do not look exactly alike. Identical twins can have differences even though they share similar genes.

A group of ants may all look the same, but like humans, there are variations among the individuals. Scientists have noticed these slight differences and observed that some individuals of species may have a survival advantage because of such a difference.

Living things will survive changes in the environment if they have a variation that allows them to obtain food, water, and shelter. If the survivors are able to reproduce, they may pass on the successful trait to their offspring. Variations are important for the survival of a species.

BIG IDEA CHECK-UP 3.8 (For answer, see page 155.)

Multiple Choice (Circle the correct answer.)

1. The birds shown below all belong to the same group. Which birds will have a survival advantage if food supply is mostly small seeds?

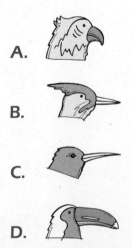

A.

B.

C.

D.

Check your answer before going on to the next topic.

3.9 LIFE CYCLES OF ANIMALS

Animals grow into adults and reproduce in order to continue the species. Not all species go through the same stages of growth and development, but all animals begin with a fertilized egg. The fertile egg contains genes from the male parent and the female parent.

Some animals such as fish are born with the appearance of smaller versions of the adult animal. Other animals such as frogs begin life looking nothing like the adult they will eventually become.

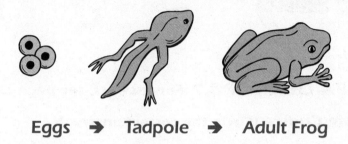

Eggs ➜ Tadpole ➜ Adult Frog

Insects may go through incomplete change or complete change. Insects that hatch looking very much like tiny versions of the adult are said to undergo incomplete change. Insects that hatch as a worm-like larva, spin a cocoon or produce a chrysalis in which they become a pupa, and finally emerge as an adult undergo complete change.

Young grasshoppers appear to be small adults, but they lack wings. As a grasshopper nymph grows, it sheds its hard outer covering, replacing it with a new larger outer body. When the nymph reaches the final adult stage, it will be larger and have wings.

Butterflies undergo complete change. They hatch with a worm-like appearance. The larva is called a caterpillar. It eats and grows. When the time is right, the caterpillar will produce a hard protective covering called a chrysalis. Inside the chrysalis, the cells of the caterpillar are rearranged into the body, legs, and wings of an adult butterfly. When the change is complete, the chrysalis breaks open and the adult butterfly emerges.

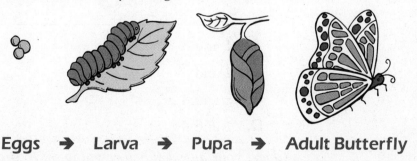

Eggs ➜ Larva ➜ Pupa ➜ Adult Butterfly

Animals with backbones that live on land develop from covered eggs or are born live. Reptiles such as turtles have eggs with a leather-like covering. Bird eggs have a hard outer shell of calcium. Mammals such as cows give birth to live young.

BIG IDEA CHECK-UP 3.9 (For answers, see page 155.)

Multiple Choice (Circle the correct answer.)

1. Which of the following is the correct order for the life cycle of a grasshopper?

    A. Egg, nymph, adult

    B. Egg, larva, pupa, adult

    C. Egg, embryo, hatchling, adult

    D. Egg, caterpillar, chrysalis, adult

2. An embryo is a stage of development. Look carefully at the animal embryos. Which two animals are most closely related?

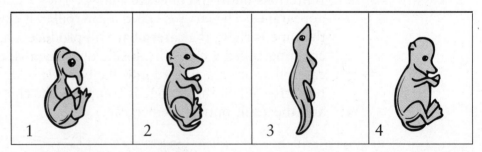

    A. 1 and 3

    B. 2 and 3

    C. 2 and 4

    D. 1 and 4

Check your answers before going on to the next topic.

PHYSICAL SCIENCE: CHEMISTRY

CHEMISTRY CONCEPT MAP

Matter
has physical properties that can be used for classification, for example,

Metals	Nonmetals
Hard	Brittle
Shiny	Dull
Solid except for mercury	Solid, liquid, or gas
Conduct electricity	Do not conduct electricity
Some attracted by magnets	Not attracted by magnets
Dense	Usually not dense

matter can be

either	or
Pure Substances that have unique physical and chemical properties that can be used for identification	**Mixtures of Substances** that can be separated based on their physical properties

can	or can
Undergo a **physical change** The appearance may change, but no new substance forms	Interact in a **chemical change** to make a new substance with new physical and chemical properties

for example	for example
Ice melts to become liquid water, both substances are made of the same type of matter but one is a solid and the other a liquid	Baking soda reacts with vinegar to form new substances

Phase changes occur at specific temperatures unique to a substance	
for example	
Gas	Water vapor
Boiling Point	100°C or 212°F
Liquid	Water
Freezing Point	0°C or 32°F
Solid	Ice

4.1 PHYSICAL PROPERTIES OF MATTER

THE BIG IDEA

Matter has physical properties, some of which can be observed with the human eye and some of which require a magnifying glass.

Matter is anything that takes up space and has mass. Matter can be densely packed together as a solid, more fluid as a liquid, or very loosely organized as a gas. Matter has physical properties that can be used to classify the type of matter. Some physical properties of matter are texture, color, shininess, hardness, and density.

Two common categories of matter are metals and nonmetals. The table compares the characteristics of metals and nonmetals that can be used for identification.

Metals	Nonmetals
Hard	Brittle
Shiny	Dull
Solid except for mercury	Solid, liquid, or gas
Conduct electricity	Do not conduct electricity
Some attracted by magnets	Not attracted by magnets
Dense	Less dense
Examples: copper and iron	Examples: sulfur and oxygen

A sample of matter may be large enough that the properties of shape, color, and texture can be seen with the human eye, for example, a penny. You can easily see that a new penny is shiny and reddish orange in color. You can feel that it is a hard substance. When you drop a penny into water it sinks. These are all physical properties of matter that would help you conclude that a penny is made of matter and that it is probably an example of a metal.

Some particles of matter may be so tiny that special tools are needed to see the details of its physical properties. For example, chemicals in table salt combine to form a specific shape. The grains of table salt are usually very small. When you pour table salt onto a dark surface, it may look like tiny grains of sand. You would need a magnifying class to see the tiny individual cubes of salt. Salt is a solid. Solids have a definite shape and volume. (See figure at top of page 61.)

Water is made of matter, but it is transparent. Light passes through the liquid. You can see the effect of water on light by placing a pencil in a glass of water. The pencil will appear to bend. Water molecules cause the light to bend. Even substances that are transparent are made of matter. Water is a liquid. It takes the shape of its container, but occupies a definite amount of space.

Water can cause
light to bend.

Liquids take the shape of
the container, but cannot be
made to fit a smaller space.

Some matter is so spread out that that it cannot be seen at all. For example, the oxygen that you need to survive is in air, but you cannot see it. Oxygen is a gas. Gases have no definite volume and no definite shape. The molecules of gas spread out as they are heated and come together when they are cooled. Gases can be pressed into a smaller space.

You can prove that oxygen is in a sample of air by using a lit candle under a jar of air that is inverted in a pan of water. The candle will burn until all the oxygen is consumed. Notice in the illustration that the water level rises inside the jar. When fire removed the oxygen from the air, the water filled the space it left behind. Even matter we cannot see takes up space.

Copper, oxygen, salt, and water are examples of pure substances. They are all made of one type of matter. Pure substances have properties that can be used for identification. These properties can also be used to separate them from a mixture of substances that do not interact to make a new substance.

For example, salt can dissolve in a cup of warm water. Salt water is a mixture that can be easily separated. Pour the salt and water mixture into a shallow pan. Wait a few days. The water will evaporate into the air leaving behind salt crystals. Mixtures are two or more substances that can be easily separated from each other.

BIG IDEA CHECK-UP 4.1 (For answers, see page 155.)

Multiple Choice (Circle the correct answer.)

1. Air is invisible. Which of the following would show that air has mass?

   A. An inflated balloon is heavier than a deflated balloon.

   B. An inflated balloon floats on water.

   C. An inflated balloon is larger than a deflated balloon.

   D. An inflated balloon will become smaller in a cold place.

2. Iron filings have been mixed with sand. Which tool would quickly separate the two substances?

   A. Tweezers

   B. Sifter

   C. Magnet

   D. Magnifying glass

3. A helium balloon expands in a warm room. What type of matter is helium?

   A. A gas

   B. A solid

   C. A liquid

   D. A vapor

Check your answers before going on to the next topic.

4.2 PHYSICAL CHANGES

A physical change occurs when the appearance of a pure substance changes, but the type of matter does not change. When salt dissolves in water it undergoes a physical change because the particles of salt become so small they fit into spaces in the water. However, the salt is still salt.

A major type of physical change is the **phase change**. There are three main phases of matter: solid, liquid, and gas. Matter is made of tiny particles that no human has ever seen. When matter is a solid, the particles are packed tightly together. When matter is a liquid, the particles are more spread out, but still linked together. Matter is a gas when the particles are widely scattered.

Let's use your class as a model. When all the students are seated at desks, the class has a definite shape and takes up a fixed amount of space. It seems as though the particles of matter (the students) are not moving. However, look more closely, each student is moving in his or her seat. That's how the particles of matter in a solid behave. They are not moving around, but they vibrate.

Substances form solids at their freezing points. They form liquids at the melting point. Liquids take up a fixed amount of space, but take the shape of their container.

When your class forms a line and moves through the hallway, the class has no definite shape, but it does take up a certain amount of space. The particles are moving much faster than they were in a solid, but they still have some organization in relationship to each other. Your line moving through the hallway is a liquid being poured from one container to another.

Liquids become a gas at the boiling point. Some liquids can become a gas by evaporation that occurs below the boiling point. Gas particles are sensitive to temperature. They move faster when they are heated and can expand. They slow down when cooled. If cooled enough, a gas can become a liquid.

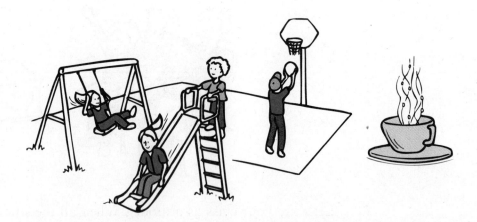

Your class during recess is a gas. The students are spread out, moving all over the playground, and sometimes colliding with each other. Excited gas molecules move very fast. Gases have no fixed volume and no fixed shape.

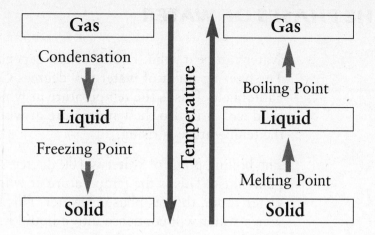

Whether a substance is a solid, liquid, or gas depends on the temperature. The temperature at which a phase change occurs can be used to identify a pure substance and used to separate mixtures.

BIG IDEA CHECK-UP 4.2 (For answers, see page 155.)

Multiple Choice (Circle the correct answers.)

1. Water freezes at 32 degrees Fahrenheit. What phase of matter would water be in if the temperature is 40 degrees Fahrenheit?

   A. Solid

   B. Liquid

   C. Gas

   D. Vapor

2. Which of the following does the diagram shown above demonstrate?

   A. Liquids are made of matter

   B. Liquids have a definite shape

   C. Liquids have a definite volume

   D. Liquids expand when heated

Check your answers before going on to the next topic.

4.3 THE PHASES OF WATER

Water can be a solid, liquid, or gas depending on the temperature. The freezing point of water is 0 degrees Celsius or 32 degrees Fahrenheit. This is the temperature at which liquid water becomes solid ice. It is also the temperature at which ice will begin to melt if the temperature is rising.

The boiling point of water is 100 degrees Celsius or 212 degrees Fahrenheit. This is the temperature at which liquid water becomes water vapor, the gas phase of water. This is also the temperature at which water will condense into a liquid if the temperature is cooling.

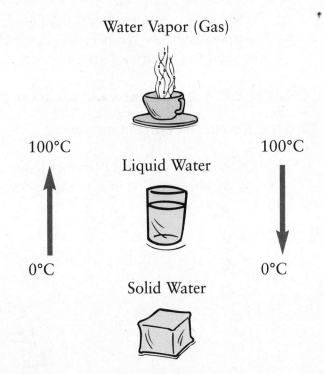

Water Vapor (Gas)

100°C   100°C

Liquid Water

0°C   0°C

Solid Water

Even though the physical characteristics of water change as the temperature changes, the chemicals that make up water do not change. The individual particles of water spread out as they warm up and come together when they cool down. Most substances lose volume when they freeze, but water expands as it becomes a solid. This is because of the chemical structure of water.

In nature water will become a vapor at temperatures much lower than the boiling point. This is the result of air moving across the surface of the water and sunlight striking the water surface.

You may have experienced this during the summer when you came out of the swimming pool. The air and sun dried your bathing suit. Perhaps you felt a chill as the water dried from your skin. This is called **evaporation**.

BIG IDEA CHECK-UP 4.3 (For answers, see page 156.)

Multiple Choice (Circle the correct answer.)

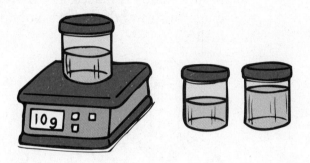

Carla placed water into a container. She weighed the container and water. The mass was 10 grams. She froze the water and then weighed the ice.

1. What is the mass of the frozen water?

    A. The mass is greater than 10 grams because the ice is larger.

    B. The mass is less than 10 grams because the ice is less dense.

    C. The mass will be the same because no water was added or lost.

    D. It is not possible to predict the mass of the frozen water.

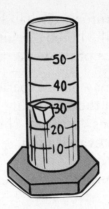

2. If an ice cube is placed in a graduated cylinder of water, which diagram shows the correct volume of water after the ice cube melts?

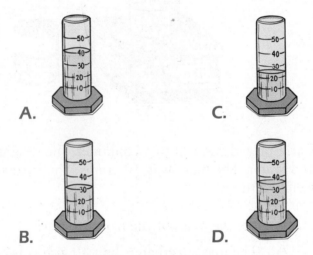

A.

C.

B.

D.

4.4 CHEMICAL PROPERTIES OF MATTER

Matter is made of building blocks that are so small no human has ever seen them. We know that they exist because sometimes when two different substances mix they react to form new substances.

Substances have chemical properties that allow scientists to predict how they will react. One property is called pH. When red cabbage is boiled, it makes a dye that changes color, depending on the pH of a substance. We call such a chemical an **indicator**.

In vinegar, the red cabbage indicator turns red. In water, the red cabbage indicator turns purple-blue. In a mixture of water and baking soda, the red cabbage indicator turns green-yellow.

Vinegar tastes sour and causes a slight burning sensation on human skin. Water has no taste and feels wet on human skin. Baking soda is a powder that tastes bitter when mixed into water. The mixture will also feel slippery, like soap.

Sour taste, burning sensation, and a pH that turns red cabbage juice red are chemical properties of acids. No taste, wet feel, and a pH that turns red cabbage juice purple-blue are chemical properties of a neutral substance. Bitter taste, slippery feel, and a pH that turns red cabbage juice green-yellow are chemical properties of a base or alkaline substance.

4.5 CHEMICAL CHANGES

Chemical changes result in the formation of new substances. Think of the building blocks of matter as letters of an alphabet. The letters can be arranged in patterns that have meaning. If the letters are rearranged, then the patterns change meaning or lose meaning. Chemical reactions occur when the rearrangement of the letters will result in new words that have a different meaning.

$$Mat + Cap \rightarrow Map + Cat$$

Mat means a small rug. Cap is a hat. They represent our starting substances. In our chemical reaction the letters mix and change places making two new words: map and cat. A map is not a mat, and a cat is not a cap. No letters were lost; they were simply rearranged. That is how a chemical change makes a new substance. The building blocks are mixed up and put back together to make something new.

If vinegar and baking soda are mixed together, bubbles form. This is evidence that a chemical reaction is taking place.

If red cabbage juice is added to the vinegar, the vinegar will have a red color. If baking soda is added to the vinegar, the color of the mixture will turn purple-blue. If more baking soda is added, the mixture may turn green. This is another sign that a chemical reaction has taken place.

Experimental Procedure

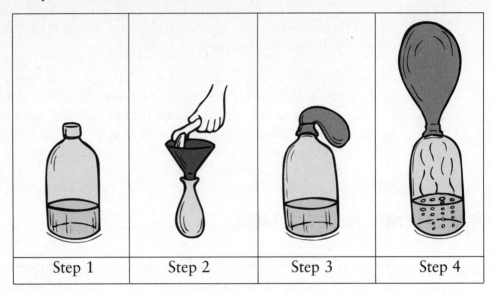

| Step 1 | Step 2 | Step 3 | Step 4 |

Jamal followed the steps shown in the experimental procedure. He poured $1/2$ cup of vinegar into an empty, clean 1-liter soda bottle. Then he measured 1 tablespoon of baking soda. Using a funnel, Jamal poured the baking soda into a balloon. He then carefully placed the balloon onto the neck of the soda bottle so that it formed a tight seal. Finally he tipped the balloon so that the baking soda dropped into the vinegar.

Jamal observed that the vinegar began to bubble. Next he noticed that the balloon inflated. When he placed his hands on the balloon, it felt warm. When he placed his hands around the bottom of the bottle, it felt cool. Changes in temperature show that energy has been gained or lost during a chemical reaction.

BIG IDEA CHECK-UP 4.5 (For answers, see page 156.)

Open-Ended Questions

Randy used red cabbage juice indicator to test for chemical reactions between vinegar and baking soda and between salt and water.

Substance	Color in Red Cabbage Juice
Vinegar	Red
Vinegar and baking soda	Blue
Water	Blue
Water and salt	Blue

1. Did a chemical reaction take place between vinegar and baking soda? Explain your answer.

   _____

   _____

2. Did a chemical reaction take place between salt and water? Explain your answer.

   _____

   _____

Check your answers before going on to the next topic.

PHYSICAL SCIENCE: PHYSICS

PHYSICS CONCEPT MAP

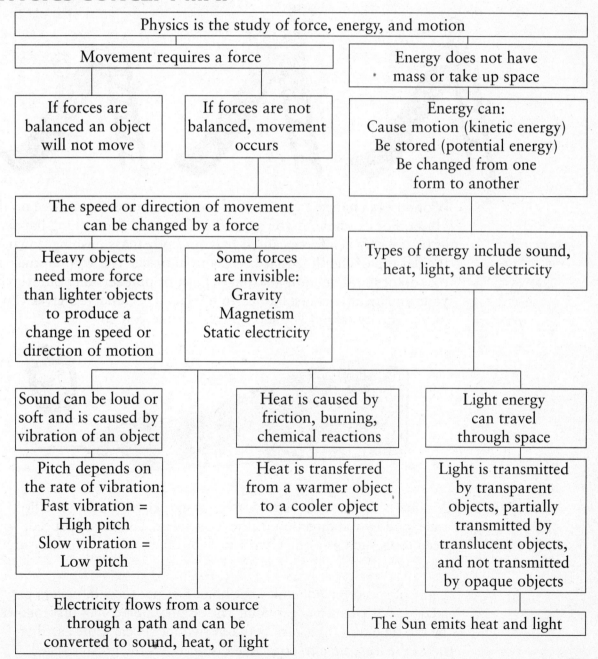

Physics is the study of force, energy, and motion

Movement requires a force

Energy does not have mass or take up space

If forces are balanced an object will not move

If forces are not balanced, movement occurs

Energy can:
Cause motion (kinetic energy)
Be stored (potential energy)
Be changed from one form to another

The speed or direction of movement can be changed by a force

Heavy objects need more force than lighter objects to produce a change in speed or direction of motion

Some forces are invisible:
Gravity
Magnetism
Static electricity

Types of energy include sound, heat, light, and electricity

Sound can be loud or soft and is caused by vibration of an object

Heat is caused by friction, burning, chemical reactions

Light energy can travel through space

Pitch depends on the rate of vibration:
Fast vibration = High pitch
Slow vibration = Low pitch

Heat is transferred from a warmer object to a cooler object

Light is transmitted by transparent objects, partially transmitted by translucent objects, and not transmitted by opaque objects

Electricity flows from a source through a path and can be converted to sound, heat, or light

The Sun emits heat and light

5.1 MOTION AND FORCES

> **THE BIG IDEA**
>
> A push or a pull is needed for motion. Another push or pull can change the speed or direction of an object in motion.

Motion is a change in position of an object caused by a force on the object. For example, if you sit at the top of a slide, nothing happens at first. There is a force caused by your body mass pushing down on the flat top of the slide that keeps you in place. You will be stuck at the top of the slide until a stronger push or pull acts on you. Maybe your impatient friend standing on the ladder behind you gives you a push. Gravity takes over and down the slide you go.

Motion is described by speed, the distance covered in a specific time, and by the direction of the movement. To measure speed, you need tools such as a ruler, yardstick, or tape measure. To measure time, you need a clock or watch.

Units of speed include miles per hour, feet per second, meters per second, and kilometers per hour. The distance unit comes first followed by the time unit. In a race between two model cars, the car that takes the least time to cover the distance has the greater speed.

Speed and direction can be changed by a push or pull on the moving object. During a soccer game, one player might kick the ball toward the goal, but before it can cross the goal line, a player on the other team kicks the ball from the side. The direction of the ball will change because a new force has been applied to the ball. Depending on the force of the kick, the speed may also change.

It takes more force to change the speed and direction of a heavy object compared to a lighter object. Think about pulling a wagon. An empty wagon is easy to move. A wagon filled with sand would take a much stronger pull force to cause movement.

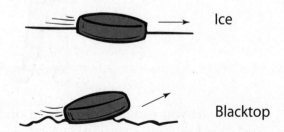

The type of surface can affect the rate and direction of motion. A hockey puck struck by a hockey stick on ice will move very fast and straight over smooth ice. The same puck struck by a hockey stick on blacktop may bounce, slow down, or change direction because of bumps in the surface. (See figure at the bottom of page 75.)

Some forces can change the motion of an object without actually touching the object. Magnetism, gravity, and static electricity are examples of forces that can push or pull an object from a distance.

Magnets have two ends: a north pole and a south pole. North attracts south, but north repels north. In other words, unlike poles pull toward each other, and like poles push each other away.

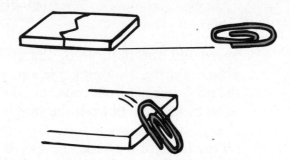

A magnet can be used to pull a metal paperclip across a desktop. If pulled to the edge of the desk, the paperclip will fall to the floor. Gravity pulls the paperclip toward the floor.

Gravity is a force on objects caused by the enormous mass of the Earth compared to the mass of objects on the Earth. When you jump up, gravity brings you down.

If you rub a latex balloon on your shirt during the winter, it will build up a static electrical charge. Hold the balloon over some small scraps of paper. The static electricity will cause the bits of paper to move up toward the balloon. The paper and the balloon are made of materials that do not conduct electricity.

Examples of Static Electricity

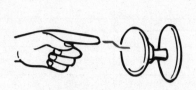  

During the winter, when the air is dry, you may have walked across a carpet and touched a metal doorknob. Ouch. A spark jumped between your fingers and the doorknob. This is static electricity that built up as you collected charges from the carpet.

Another example of static electricity can be demonstrated using a plastic comb. If both your hair and the air are dry, the comb will cause your hair to stand on end because of a transfer of charges from your hair to the comb.

Lightning is a strong static electric discharge between clouds and the ground. This is an extremely dangerous type of static electricity.

BIG IDEA CHECK-UP 5.1 (For answers, see page 157.)

Multiple Choice (Circle the correct answer.)

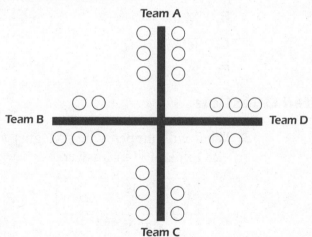

1. The diagram shows a game of 4-way tug-of-war. The circles represent students on each team. If each individual student pulls with the same force, which team will cause the flag to move away from the center?

   A. Team A   **C.** Team C

   B. Team B   **D.** Team D

Results of 1-mile Run

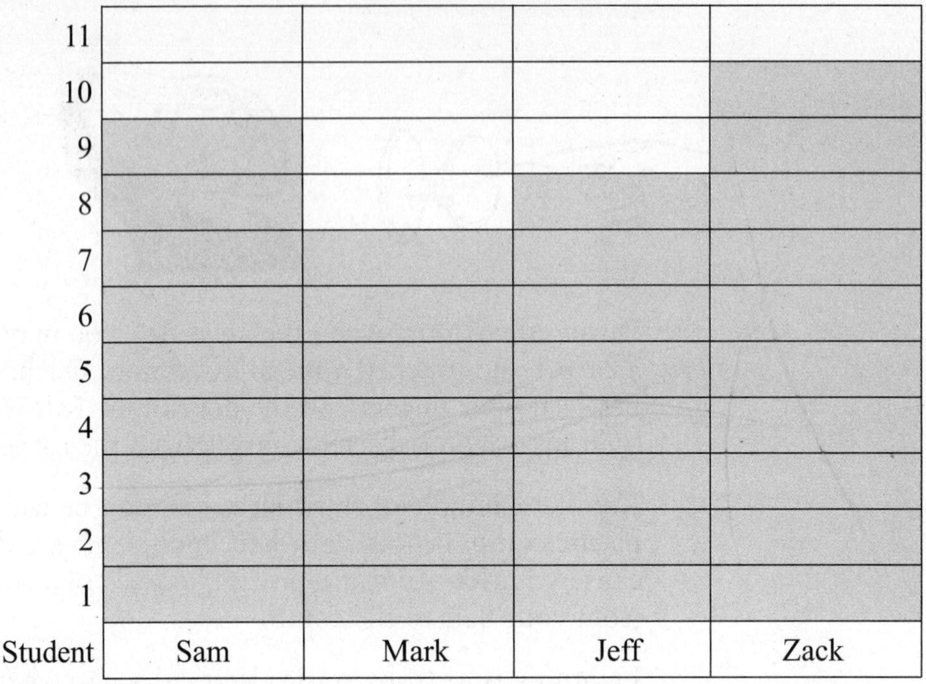

Student	Sam	Mark	Jeff	Zack

(Y-axis values: 11, 10, 9, 8, 7, 6, 5, 4, 3, 2, 1)

2. The graph shows the time it took for each student to run 1 mile. Which student had the fastest speed?

   A. Sam

   B. Mark

   C. Jeff

   D. Zack

Open-Ended Question

3. What will happen when Magnet #1 is moved closer to Magnet #2? Explain your answer.

S	N		N	S

Magnet #1   Magnet #2

_____

_____

_____

_____

Check your answers before going on to the next topic.

5.2 ENERGY

THE BIG IDEA

Energy can cause motion and be stored. Energy can also be changed from one type to another.

Energy does not have mass and does not take up space. It is needed in order for motion to occur or work to be done. Forces need energy.

Kinetic energy is the energy of motion. When you blow on a pinwheel, the pinwheel spins. The moving air has kinetic energy. When the air strikes the pinwheel, the kinetic energy is changed into mechanical energy.

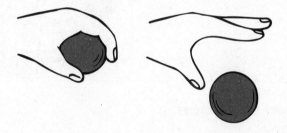

Hold a ball in your hand so that it faces the floor. The ball has stored energy. It has the potential to move. The potential energy is changed to kinetic energy when you open your hand and the ball begins to fall toward the floor.

Other forms of energy are sound, heat, light, and electricity. The Sun is the main source of heat and light for Earth.

Open Switch
No Energy Flow

Closed Switch Energy Flow
Electricity ➜ Light + Heat

Close the switch. Electricity flows through wires to a light bulb. The light bulb glows and becomes warm. Energy has been transformed from one type to another. (See figure on bottom of page 79.)

BIG IDEA CHECK-UP 5.2 (For answers, see page 158.)

Multiple Choice (Circle the correct answer.)

1. What type of energy does a falling ball have?

   A. Potential energy

   B. Gravity

   C. Kinetic energy

   D. Friction

Open-Ended Question

2. Describe the energy transformation shown in the diagram.

   _____

   _____

   _____

   _____

   _____

Check your answers before going on to the next topic.

5.3 SOUND

Sound travels in waves through air, water, or a solid object. Vibrations of matter cause sound. Humans hear sound because the energy causes special structures in the ear to vibrate.

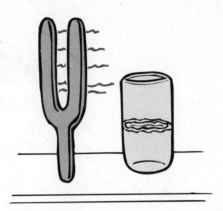

You cannot see sound waves directly, but you can observe their effects. Strike a tuning fork. Place it on a table near a glass of water. The water will begin to vibrate.

Sound has pitch. Slow vibrations make low-pitched sounds. Fast vibrations make high-pitched sounds.

Guitars have thin and thick strings. The thin strings vibrate faster than the thick strings. The thin strings produce high-pitched sounds while the thick strings make low-pitched sounds.

A guitarist can change the pitch of the string by pressing a finger onto the string. This shortens the string causing the string to vibrate faster.

BIG IDEA CHECK-UP 5.3 (For answers, see page 159.)

Multiple Choice (Circle the correct answer.)

Chelsea stretched four rubber bands around a shoebox lid. Each rubber band was a different thickness. Use this information and the diagram to answer questions 1 and 2.

A   B   C   D

1. Which rubber band shown will produce the lowest pitched sound?

   A. Rubber band A

   B. Rubber band B

   C. Rubber band C

   D. Rubber band D

Open-Ended Question

2. If Chelsea places sand in the bottom of the shoebox lid, what will happen when she produces sound from the rubber bands?

   _____

   _____

   _____

   _____

   _____

   _____

   _____

Check your answers before going on to the next topic.

5.4 HEAT

THE BIG IDEA

Heat is a form of energy that flows from an area of higher heat to an area of less heat.

Heat is a form of energy that can travel through space. Heat energy from the sun warms Earth's surface. Some of the heat is trapped in Earth's atmosphere. Some escapes into space. A greenhouse works like the atmosphere to trap heat from the sun.

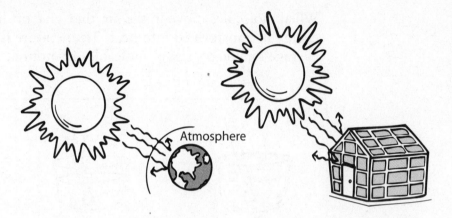

Heat can also result from energy transformations. When electricity flows through a metal that slows down the flow, some of the electricity is transformed into heat.

Some metals are good conductors of heat. Hot water flowing through a metal radiator will warm the surrounding air.

Other materials act as insulators and stop the loss of heat. Pipes running from the hot water boiler to the radiator will be covered with an insulator to stop the loss of heat. Insulators act like blankets to trap heat.

Friction is another source of heat energy. When two substances rub against each other, their kinetic energy can be transformed into heat. When you rub your hands together, you can feel the heat.

When molecules move in the air, they give off heat. Their kinetic energy is transformed into heat. Temperature is a measure of the average kinetic energy of molecules in motion.

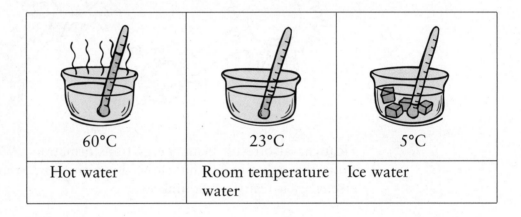

60°C	23°C	5°C
Hot water	Room temperature water	Ice water

Hot and cold are relative terms. If you place your hand in ice water and then in room temperature water, the water at room temperature will feel warm. If you place your hand in hot water, then in room temperature water, the water at room temperature will feel cold. The temperature of the water does not change even though your perception changes. A thermometer provides a measurement of the energy of a system or an object.

Heat flows from where there is more energy to where there is less energy. Hot chocolate loses heat into the air and becomes cool enough to drink. If left sitting long enough, the mug of hot chocolate would have a temperature equal to the air temperature. Heat stops flowing when both objects reach the same temperature.

BIG IDEA CHECK-UP 5.4 (For answers, see page 159.)

Kevin measured the temperature of ice cubes in a glass and recorded the temperature. Then he measured the temperature of water in another glass and recorded the temperature. Finally he poured the water over the ice cubes and measured the temperature of the mixture.

Kevin continued to measure the temperature at 15 minutes, 30 minutes, and 60 minutes. That data and observations are shown in the table. Use the information to answer the questions.

Setup

| Ice cubes 0°C | Water 23°C | Water + Ice Cubes 22°C |

Data and Observations of a Glass of Water and Ice

Time of Observation	Temperature	Observation
15 minutes	15°C	Ice cubes are smaller
30 minutes	10°C	One small piece of ice left
60 minutes	20°C	Ice is gone, water droplets on the side of the glass

Air temperature was 23°C at the beginning and end of the experiment.

Multiple Choice (Circle the correct answer.)

1. Why did the water become cooler after 30 minutes?
   A. The ice transferred coldness to the water.
   B. The ice transferred coldness to the air.
   C. The water transferred heat to the air.
   D. The water transferred heat to the ice.

2. Why did the water become warmer at 60 minutes?

   A. The water gave up heat to air as water evaporated.

   B. The air was hotter than the water in the glass.

   C. The ice had melted and was no longer giving up coldness.

   D. The glass was heating up in the warm air.

Open-Minded Question

3. What could Kevin do to keep the temperature of the water at 10°C for a longer period of time?

_____

_____

_____

_____

_____

Check your answers before going on to the next topic.

5.5 LIGHT

THE BIG IDEA
Light is a form of energy that can be transmitted by transparent objects, partially transmitted by translucent objects, and reflected by opaque objects.

Light is a form of energy that can travel through space. Sometimes light acts as a wave and sometimes as though it were made of invisible particles that can pass through transparent objects, partially pass through translucent objects, and bounce off opaque objects.

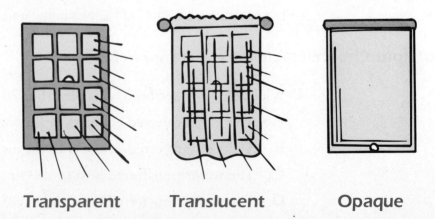

Transparent **Translucent** **Opaque**

A glass window allows light to move through the window. Glass is transparent.

A sheer curtain will cause some of the light to be reflected back while allowing some light to pass through. The curtain is translucent.

A shade blocks the light by reflecting it back through the window. The shade is opaque.

Some substances such as water will bend light. As light passes from air through the glass, it slows down a little. As the light passes through water, it slows even more and changes direction. The light moves faster as it exits the glass and enters air. This is called **refraction**. If you place a pencil in a glass of water, the pencil will appear to bend because of this effect.

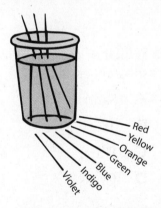

Catch a sunbeam with a glass of water. A rainbow will appear. Light contains the colors that you see. Objects reflect the parts of light that make up their colors and absorb the rest. For example, grass appears green because it is reflecting back green. You can only see reflected light.

Natural sources of light are stars. Our closest star is the Sun. Electricity can be converted to light. Some animals produce light by chemical means, for example, lightning bugs.

BIG IDEA CHECK-UP 5.5 (For answers, see page 161.)

Multiple Choice (Circle the correct answer.)

Use the illustration below to help you answer questions 1 and 2.

1. Which of the following is an example of a transparent substance?

    A. Sand

    B. Tree

    C. Water

    D. Grass

Open-Ended Question

2. Explain why there is a shadow in the diagram.

    _____

    _____

    _____

Check your answers before going on to the next topic.

5.6 ELECTRICITY

THE BIG IDEA

Electricity is a form of energy that can produce a magnetic field and be transformed into heat and light.

Electricity is a form of energy that is a flow of electrons. In order to move, electricity needs a source of electrons, a path along which the electrons can flow, and something that uses the electrons.

The diagram shows a circuit. The copper penny, lemon, and nail are the source of the electrons. Like most battery cells, this is chemical energy.

The connecting wire is the path of the electrons. Metal is a good conductor of electricity.

The light bulb uses the electrons to make light. The light bulb contains a thin special metal wire that causes the electricity to slow down. When this happens, the electricity is converted to light.

Electricity can travel through materials called conductors. Metals are considered good conductors of electricity.

Plastic is an insulator. Insulators do not let electrons flow through them. Wires in a circuit are usually covered with plastic. The plastic keeps electrons from being lost and prevents touching the wires from becoming a shocking experience.

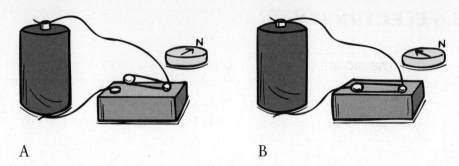

A   B

When electricity passes through a wire, it produces a magnetic field. In diagram A, the wire is only connected to one side of the battery cell. No electrons are flowing through the wire. The compass needle is pointing North in response to Earth's magnetic field.

In diagram B, the wire is connected to both ends of the battery cell. Electrons are flowing through the wire. The compass needle moves, showing that the wire has its own magnetic field.

BIG IDEA CHECK-UP 5.6 (For answers, see page 161.)

Multiple Choice (Circle the correct answer.)

Use the diagram of a flashlight to answer questions 1 and 2.

1. What is the source of electricity to power the flashlight?
   A. Light bulb   C. Switch
   B. Wires   D. Battery cells

2. Which of the following shows the correct energy transformation when the switch is turned on?
   A. kinetic ➜ electrical ➜ light
   B. chemical ➜ electrical ➜ light
   C. magnetic ➜ electrical ➜ light
   D. potential ➜ electrical ➜ light

Check your answers before going on to the next topic.

Chapter 6

EARTH SCIENCE

EARTH SCIENCE CONCEPT MAP

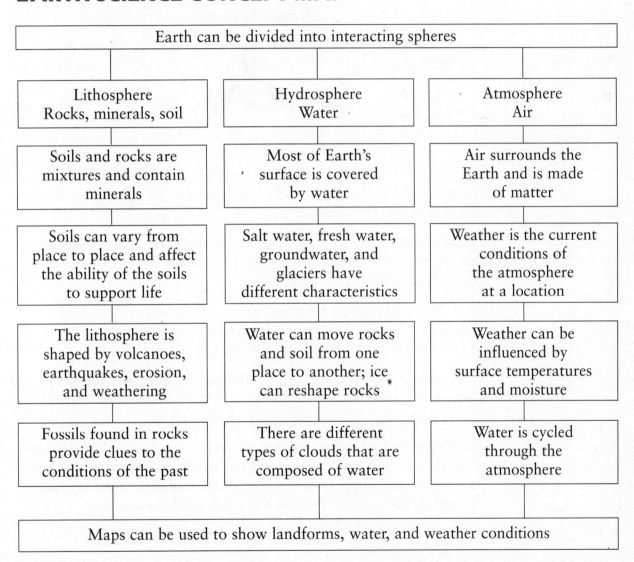

Earth can be divided into interacting spheres

Lithosphere Rocks, minerals, soil	Hydrosphere Water	Atmosphere Air
Soils and rocks are mixtures and contain minerals	Most of Earth's surface is covered by water	Air surrounds the Earth and is made of matter
Soils can vary from place to place and affect the ability of the soils to support life	Salt water, fresh water, groundwater, and glaciers have different characteristics	Weather is the current conditions of the atmosphere at a location
The lithosphere is shaped by volcanoes, earthquakes, erosion, and weathering	Water can move rocks and soil from one place to another; ice can reshape rocks	Weather can be influenced by surface temperatures and moisture
Fossils found in rocks provide clues to the conditions of the past	There are different types of clouds that are composed of water	Water is cycled through the atmosphere

Maps can be used to show landforms, water, and weather conditions

6.1 ROCKS, MINERALS, AND SOIL

THE BIG IDEA
Rocks can be classified based on their characteristics.

6.1.A ROCKS

The surface of the Earth is called the **lithosphere**. Litho means rock. Rocks are made of minerals and other substances. Some rocks contain chemicals from deep within the earth. Others are formed from surface materials.

Igneous Rocks

Rocks from Cooled Magma		Rocks from Cooled Lava
· Granite		Basalt
Diabase		Pumice

Some rocks form when **magma** deep inside the Earth cools. Such rocks will have large crystals because they cooled over millions or billions of years.

When magma rises to the surface it is called **lava**. When rocks from lava form at the surface, they often have small crystals from the short cooling time.

Both lava and magma rocks are called **igneous** *(ig-nee-us)* rocks. Igneous rocks appear to be made of one substance. Sometimes igneous rocks have visible crystals and are often dark in color.

This type of rock is dense. An exception is pumice that forms as lava cools in the air. Pumice floats on water! Basalt, diabase, and granite are common igneous rocks of New Jersey.

Rocks can break up into sediments. Shells from dead ocean animals dissolve into salt water. Over time the sediments become chemically combined into a new type of rock. Many **sedimentary** rocks form in oceans and contain carbonate. They will bubble in vinegar.

Sedimentary Rocks

Rocks from Sediments
Sandstone
Limestone

Mixing sand, chalk, or tiny pebbles into plaster of Paris is one way to observe how sedimentary rocks form. The sand grains and rock chips can sometimes be seen with a magnifying glass. Sandstone and limestone are common sedimentary rocks found in New Jersey.

Metamorphic Rocks

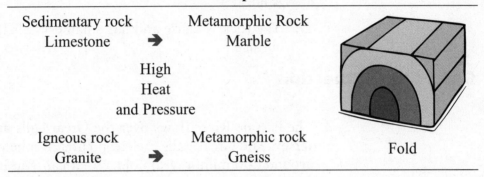

Sedimentary rock		Metamorphic Rock
Limestone	➜	Marble
	High	
	Heat	
	and Pressure	
Igneous rock		Metamorphic rock
Granite	➜	Gneiss

Fold

A third type of rock forms when other rocks are changed by high heat and pressure inside the Earth. Metamorphosis *(met-a-more-fo-sis)* is a word that means change. **Metamorphic** rocks are changed rocks. Because they form from high pressure and heat, they often have visible layers.

Granite is an igneous rock that can become gneiss *(nice)*. The black flecks that are scattered in some types of granite become more organized into layers of black and white stripes in gneiss.

Limestone is a sedimentary rock that can become marble. Marble does not have obvious layers. Marble will slowly dissolve in an acid such as vinegar.

BIG IDEA CHECK-UP 6.1.A (For answers, see page 161.)

Multiple Choice (Circle the correct answer.)

1. Read the rock descriptions. Which one is probably metamorphic based on the characteristics?

   A. The rock is light gray and leaves a mark when rubbed onto paper.

   B. The rock is dense and has black specks between white layers.

   C. The rock appears to be made of many smaller pebbles and sand.

   D. The rock is dense and has small crystals throughout the surface.

Open-Ended Question

2. The Passaic River flows over the Great Falls at Paterson, New Jersey. The Great Falls are steep cliffs of igneous rock that are around 200 million years old. What can you infer happened 200 million years ago to form the Great Falls?

   _____

   _____

   _____

   _____

Check your answers before going on to the next topic.

6.1.B THE ROCK CYCLE

The Rock Cycle
Igneous Rocks

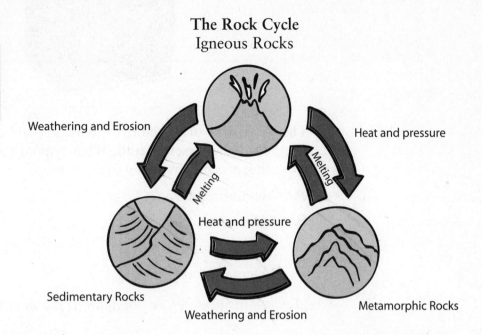

Rocks can be changed from one type to another through the rock cycle. Igneous and metamorphic rocks become sediments through the process of weathering and erosion. The sediments can undergo a chemical reaction and become sedimentary rocks.

Sedimentary rocks and igneous rocks can become metamorphic rocks under conditions of high heat and pressure caused by the movement of Earth's tectonic plates.

Both sedimentary and metamorphic rocks can be pushed deep underground where they melt and become magma. Magma and its volcanic form lava become igneous rock.

BIG IDEA CHECK-UP 6.1.B (For answers, see page 162.)

Multiple Choice (Circle the correct answer.)

1. Which process would begin to change a metamorphic rock into a sedimentary rock?

   A. Crystallization

   B. Melting

   C. Pressure

   D. Erosion

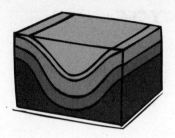

2. The movement of the North American plate caused rocks in northern New Jersey to fold. What type of rock would result from this process involving high pressure?

   A. Metamorphic rock

   B. Sedimentary rock

   C. Volcanic igneous rock

   D. Magma igneous rock

Check your answers before going on to the next topic.

6.1.C MINERALS

> **THE BIG IDEA**
>
> Rocks are composed of minerals and other substances.

Minerals are chemicals found in rocks. Quartz is an example of a mineral. The lead in your pencil is made from the mineral graphite. Some rocks called geodes contain mineral crystals, for example, amethyst, purple quartz. Graphite, quartz, and amethyst are minerals found in New Jersey.

Minerals may be used to identify where a rock was formed. Franklinite is a special combination of minerals that glow under ultraviolet light. Franklinite contains iron, zinc, and manganese and is found only in the mines of Franklin, New Jersey.

Metals such as manganese, copper, iron, and zinc may be trapped in rocks or will form bands called veins in rock. Raw metals in rock are called **ore**. Bog iron from the Pine Barrens, iron from the Highlands, and copper from the Meadowlands were important to the development of New Jersey. Magnetite and ironstone are two examples of New Jersey rocks that contain visible iron ore.

Some physical properties of minerals that are used for identification are color, luster, crystal structure, and hardness. Luster is the shininess of the mineral. Hardness is tested by scratching one mineral against another, or against a substance of known hardness. There are other tests that can be done by scientists in the laboratory. Scientists who study rocks and minerals are called geologists.

BIG IDEA CHECK-UP 6.1.C (For answers, see page 162.)

Multiple Choice (Circle the correct answer.)

Property	Mineral A	Mineral B	Mineral C	Mineral D
Color	Black	Purple	Yellow	Gray
Magnetism	Yes	No	No	No
Hardness Mineral A Scratches:	—	No	Yes	No
Mineral B Scratches:	Yes	—	Yes	Yes
Mineral C Scratches:	No	No	—	No
Mineral D Scratches:	Yes	No	Yes	—

1. Which mineral probably contains iron?

   A. Mineral A

   B. Mineral B

   C. Mineral C

   D. Mineral D

2. Which mineral is the hardest of the four?

   A. Mineral A

   B. Mineral B

   C. Mineral C

   D. Mineral D

Check your answers before going on to the next topic.

6.1.D SOIL

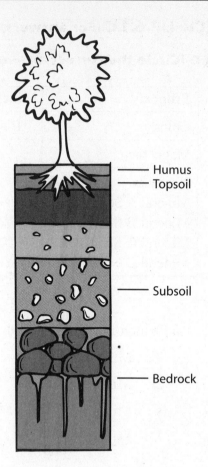

Soil is found at the outer surface of Earth's crust. Most plants use nutrients from soil for growth. When plants and animals die, they decompose. The compost from the decayed plants and animals will mix with minerals to form a rich organic surface soil called topsoil. Humans sometime add compost and fertilizer to soil to improve plant growth.

Below the topsoil is soil that is mostly minerals and very finely ground rock. Sometimes this layer is exposed to the surface. It is difficult for plants to grow in this type of soil.

Under the subsoil, there may be a layer of sand or pebbles that help soil drain. Clay is another type of deep soil. Clay does not let water flow through it. Water may collect on top of the clay forming an aquifer, an underground pool of water. Water may also run as an underground stream along the clay bed.

The final layer is bedrock. A common type of bedrock is granite, a rock that formed billions of years ago when the Earth started to cool. Diabase, basalt, and sandstone are other common bedrock types in New Jersey.

BIG IDEA CHECK-UP 6.1.D (FOR ANSWER, SEE PAGE 163.)

Open-Ended Question

Mr. Lee's class placed vegetable peels and moist clean shredded waste paper into a container with small red worms. After several weeks, a dark brown material appeared in the composter. Is this substance soil? Explain your answer.

_____

_____

_____

_____

Check your answer before going on to the next topic.

6.1.E PLANTS AND SOIL

THE BIG IDEA
Properties of soil vary from place to place and affect the ability of the soil to support life.

Soil types vary around the world. In very cold regions of the Earth, the soil may be frozen most of the year. This is called permafrost. The topsoil may thaw during the summer, but the subsoil stays frozen.

Plant life struggles in this nutrient-poor, harsh environment. Plants without true roots or with very small roots grow during the summer if the soil thaws. Lichens and some mosses may survive here. During their growing period, they will break down rock to make soil. When they die they decompose and enrich the soil.

New Jersey is not cold enough to have permafrost, but New Jersey does have a wide variety of soil types that support many different plant types.

Cacti are usually thought of as desert plants. They are adapted to living in dry sandy soil. Prickly pear cacti are common plants along the New Jersey shoreline. They can be seen along pathways to the beach at Sandy Hook. (See figure at top of page 100.)

Most grasses grow in soil deposited along flood plains. Prairie grasses survive harsh dry winters, high winds, and hot summers. Some grasses have special structures that allow them to live in salt water. The rich organic soils of New Jersey's salt marshes support cord grass and salt hay. This type of marsh can be found at Stone Harbor.

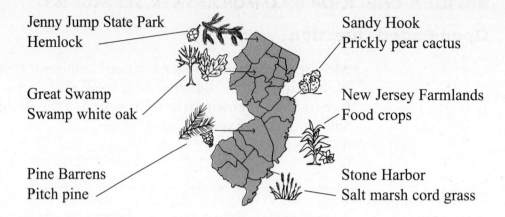

Jenny Jump State Park
Hemlock

Great Swamp
Swamp white oak

Pine Barrens
Pitch pine

Sandy Hook
Prickly pear cactus

New Jersey Farmlands
Food crops

Stone Harbor
Salt marsh cord grass

The sandy acidic soil of New Jersey's Pine Barrens supports blueberries, cranberries, and pitch pines. Unusual plants such as lady slipper and pitcher plants can grow in this soil. Pitcher plants obtain nitrogen by trapping and digesting insects.

New Jersey is the Garden State. The mixture of sandy and organic materials in the central and southern counties supports the growth of peaches, apples, corn, and many vegetables. Farmers add nutrients to the soil. Fertilizer contains the extra nitrogen and phosphate the various crops may need for growth.

A different type of soil that is a combination of materials deposited by a glacier, clay, and organic matter supports plants in the Great Swamp National Wildlife Refuge. Swamp white oak trees can grow with their roots in water.

Plants cling to rocky ledges of the ridges of Jenny Jump State Park. Eastern hemlock trees once formed thick forests in this area. Low shrubs and grasses cling to the sides of the rocky slopes. Oak, beech, and maple trees grow in deeper soil. In the valley, farmers take advantage of organic well-drained soil.

BIG IDEA CHECK-UP 6.1.E (For answers, see page 163.)

Mrs. Smith's students made four different soil mixtures from compost, sand, pebbles, and clay. They tested their four mixtures using a test for water absorption. For each soil sample, the students used the setup shown in the diagram.

The students filled each funnel halfway with a soil sample. They poured 250 ml (approximately 1 cup) of water onto the soil samples and waited 5 minutes. They recorded their observations and measurement of the amount of water in the bottom of the bottle.

Observations and Data

Soil Sample	Observations	Amount of Water in Bottom of the Bottle
A	Water dripped into the bottle quickly.	220 ml
B	Water dripped into the bottle slowly.	125 ml
C	Water stayed on top of the soil.	10 ml
D	Water ran through the soil very quickly.	248 ml

Multiple Choice (Circle the correct answer.)

1. Which soil sample in the table above is probably mostly clay?
   A. A
   B. B
   C. C
   D. D

2. Which soil mixture in the table on page 101 would be good for a cactus garden?

A. Sample A

B. Sample B

C. Sample C

D. Sample D

Open-Ended Question

3. Explain why the order of materials shown below will make a good indoor garden.

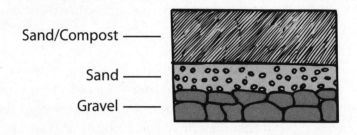

Sand/Compost ——

Sand ——

Gravel ——

_____

_____

_____

Check your answers before going on to the next topic.

6.2 FOSSILS

THE BIG IDEA
Fossils provide evidence about the plants and animals that lived long ago and the nature of the environment at that time.

Fossils provide a record of life on Earth and clues to Earth's past. Because igneous and metamorphic rocks are produced through processes of great heat and pressure, they are not likely to contain fossils. Most fossils are found in sedimentary rocks. The main types of fossils are molds, casts, imprints, petrified remains, or whole plants or animals.

Fossils are usually records of a violent death. A sudden catastrophic sandstorm in the ancient Gobi Desert left a record of dinosaurs and reptiles trapped in red sandstone that formed around them over time. Ash from the eruption of a super volcano 33 million to 34 million years ago created the petrified forests of Yellowstone Park.

Plants are soft and decompose fairly quickly. Plants are less likely than animals to be captured as fossils. Animal bones, shells, and footprints are common fossils. The remains of a plant or animal are considered a fossil if they are older than 10,000 years.

Mold Fossil

A mold is an impression of an organism. The fossil shows the details of the outer covering of the plant or animal. Molds of shells are common in New Jersey. A mold of dinosaur skin was a rare exciting find in China.

Trilobite Raised Fossil

A cast appears to be the actual plant or animal. It forms when sediments fill the mold and harden into rock. Molds of trilobites are common in the Ohio River valley.

To see how molds and casts form, press a seashell into clay. Remove the shell. The ridges and patterns of the shell will be visible in the clay. If the clay were baked to harden it, it would resemble a mold fossil. If you were to pour plaster of Paris over the clay mold and add water, a cast would form. The cast would look like the actual shell.

Petrified Wood

Petrified fossils occur when minerals replace the tissue of plants or animals. This process turns once-living tissue into rock. Petrified wood is the best known example, but animals have also been found petrified.

Bee in Amber Fossil

Tree sap can trap pollen, insects, and plant parts. The sap hardens around the plant or animal. Over time the tree sap becomes amber wrapped around a tiny treasure from the past. Scientists can study the actual organism that was trapped in amber.

The age of a fossil can be estimated by various methods. The first is done in the field when a geologist places the age of the fossil based on the age of the rock that surrounds it. The type of plant or animal that fossilized is a clue to the climate and conditions of Earth millions of years before humans.

Fossilized shark teeth, fish, and reptile bones found in Monmouth County show that this area was the bottom of the Atlantic Ocean 40 million to 25 million years ago. Dinosaur bones found in Haddonfield, New Jersey, and amber fossils of bees, maple flowers, and pollen show the diversity of life 145 million to 65 million years ago in central New Jersey. A mold of a seashell in limestone in Franklin, New Jersey, is evidence of an inland sea covering the northwestern part of New Jersey 570 million to 345 million years ago.

BIG IDEA CHECK-UP 6.2 (For answers, see page 164.)

Multiple Choice (Circle the correct answer.)

Raised Scallop Shell Fossil
60 million-year-old fossil

Questions 1 through 3 refer to the fossil shown above.

1. The fossil is an example of which of the following?

    A. Cast fossil   **C.** Mold fossil

    B. Petrified fossil   **D.** Amber fossil

2. In which of the following would this fossil most likely be found?

    A. Igneous rock

    B. Metamorphic rock

    C. Sedimentary rock

    D. Mineralized rock

3. If the fossil were found in central New Jersey, what can you conclude about this area 60 million years ago?

    A. The area was a lake.

    B. The area was an ocean.

    C. The area was a swamp.

    D. The area was a forest.

Check your answers before going on to the next topic.

6.3 AIR IN THE ATMOSPHERE

A mixture of gases and water vapor called the atmosphere surrounds Earth's surface. The atmosphere protects the Earth from the harmful waves of sunlight by causing them to bounce back into space. The atmosphere also allows necessary waves of sunlight to reach Earth's surface. These waves allow animals with eyes to see and plants to grow.

The atmosphere traps heat from the Sun like a blanket keeping the Earth at a temperature that can support life. A thinner atmosphere would make Earth too hot during the day and too cold at night. At thicker atmosphere would make Earth too hot all the time.

The atmosphere also protects Earth from objects moving through space. Very large chunks of rock, old satellites, and other objects that enter Earth's atmosphere often burn up from friction with air before they hit the ground.

We live in an ocean of air near the surface of the Earth. The layer of the atmosphere in which we live is nitrogen, oxygen, argon, and carbon dioxide. Because air is made of matter, it can press down on Earth's surface, move in response to surface heating by the sun, and cycle water.

BIG IDEA CHECK-UP 6.3 (For answer, see page 164.)

Multiple Choice (Circle the correct answer.)

Gases in Earth's Atmosphere

Argon and
Carbon dioxide

Oxygen

Nitrogen

1. Which gas is more than three-quarters of Earth's air?

A. Nitrogen

B. Oxygen

C. Argon

D. Carbon dioxide

Check your answer before going on to the next topic.

6.4 WATER IN THE ATMOSPHERE

THE BIG IDEA

Water is cycled through air by evaporation, condensation, and precipitation.

Approximately 75 percent of the surface of the Earth is covered by water. Most of the water, 97 drops out of 100, is in the oceans and seas. This type of water is salt water.

Fresh water is found in lakes, ponds, rivers, and glaciers. This type of water is called surface water. Other fresh water can be found underground in aquifers. This is called groundwater. Wells draw water from aquifers.

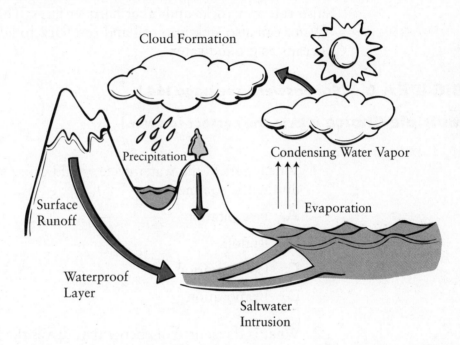

Drinking water in northern New Jersey comes mainly from surface water in reservoirs and rivers. Drinking water in southern New Jersey comes from groundwater and rivers. A large aquifer is in the center of southern New Jersey.

Life on Earth cannot exist without water. The cycling of water in the atmosphere moves water from oceans, lakes, and rivers to land. Air and sunlight start the water cycle.

As the air moves over water, the top layer of water changes to vapor. Sunlight can also bump water molecules into the air. When molecules of water move into air, the process is called **evaporation**. Water does not need to boil for evaporation to take place.

Once water molecules become vapor, they move around among the gases in air. As air rises through the lowest layer of Earth's atmosphere, it cools. As the temperature decreases, the water molecules begin to move more slowly. They begin to collect around dust particles. This is called **condensation**. If enough water molecules collect clouds may form.

When water forms droplets, they become heavy. Water will fall out of the air as precipitation in the form of rain, snow, sleet, or hail depending on the temperature of the air. **Precipitation** brings water back to the surface.

Water can fall directly back into lakes, rivers, and oceans. It can also run over rocks and other hard surfaces. This is called runoff. Water can also sink into soil and run back to lakes, rivers, and oceans as groundwater.

BIG IDEA 6.4 (For answers, see page 164.)

Multiple Choice (Circle the correct answer.)

1. Which part of the water cycle would cause soil to become dry during the summer?

   A. Precipitation

   B. Runoff

   C. Condensation

   D. Evaporation

2. What is the source of energy that drives the Earth's water cycle?

   A. Magma

   B. Sun

   C. Air

   D. Wind

Check your answers before going on to the next topic.

6.5 WEATHER

THE BIG IDEA

Weather patterns can be observed and used to predict future weather conditions.

Weather is the condition of Earth's atmosphere at a specific place and time. Factors that determine weather conditions are air temperature and the amount of water in the air. Air temperature, pressure, humidity (water in the air), and wind speed can be used to predict weather.

The tools used to study weather are shown in the table below.

Tool	What It Measures
Thermometer	Air temperature
Barometer	Air pressure
Psychrometer	Water in the air
Anemometer	Wind speed

Scientists who study weather are called meteorologists. In addition to the tools listed in the table, meteorologists use radar and satellite images to forecast weather.

WIND

Wind is the movement of air from an area of high pressure to an area of low pressure. Global winds follow predictable patterns based on the unequal heating of Earth's land and water surfaces. The rotation of the Earth is also a factor in determining the direction that air moves on Earth. (NOTE: The rotation of the Earth does not *cause* wind!)

TORNADOES AND HURRICANES

In the northern hemisphere, rising warm air rotates counterclockwise around low pressure. Falling cold air rotates clockwise around high pressure.

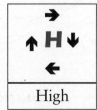

High

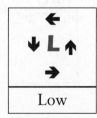

Low

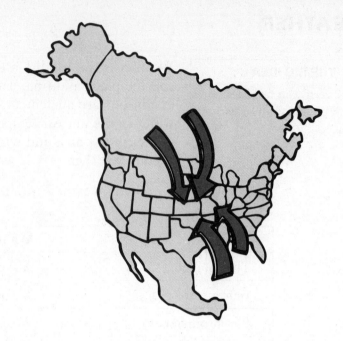

In North America, cold dry air from the arctic regions flows south toward the plains states rotating clockwise. Warm moist air flows north from the tropical regions of the Gulf of Mexico rotating counterclockwise. When the air masses collide, violent thunderstorms and tornadoes can be the result.

Thunderstorms that begin off the west coast of Africa between June and November gather strength over warm tropical Atlantic Ocean water. The rotation of the Earth causes the storms to curve to the north toward the Gulf of Mexico and east coast of the United States.

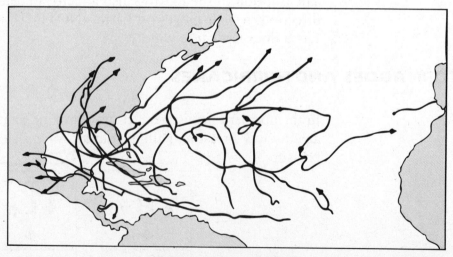

Hurricanes of 2005

If the storms develop strong rotation with high winds around a calm center, a hurricane may form. Hurricanes can cause damage to property and loss of life. Hurricane Katrina caused widespread flooding and destruction in New Orleans during 2005, the busiest hurricane season on record. There were more than 26 named storms in 2005. The hurricane season did not end until early January 2006.

New Jersey is far enough north that it is rare for a hurricane to make a direct strike on the coast. However, even the weaker tropical storm that forms as the hurricane loses speed in cooler water or after passing over land can cause heavy rainfall and flooding. Tropical storm Floyd caused damage over the entire state of New Jersey.

Flooding from heavy rains causes damage inland. Surges of seawater cause damage along the coast. The pressure of the storm pushes water ahead of it causing higher than normal tides.

NEW JERSEY WEATHER

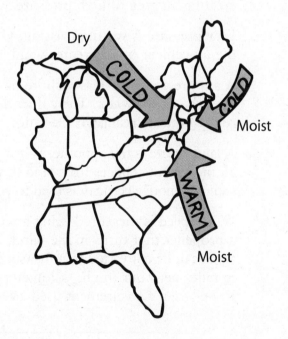

Weather tends to move from west to east. New Jersey's weather is influenced by warm moist air from the south, cold dry air from the northwest, and cold moist air from the east. Mountains in the western part of the state and the Atlantic Ocean to the east have an effect on the weather. The northern part of the state has weather similar to that in New England. The southern part of the state has weather conditions more similar to southern Atlantic states.

LOCAL WINDS

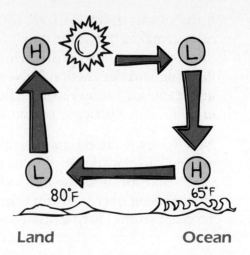

Land   Ocean

Air flows from an area of high pressure to an area of low pressure. When air at the surface of the Earth heats, it rises. As the air molecules leave, the air pressure decreases. Air rushes in from an area of high pressure to fill the space. When air rises into the atmosphere, it cools and becomes dense. Dense air is heavy and falls back to Earth creating an area of high pressure.

Low pressure is warm air rising. High pressure is cold air falling. All the movement of air creates wind at the surface.

A sea breeze at the Jersey shore happens when the land is hotter than the ocean water. The colder air over the ocean moves in to replace the air rising off the land.

A land breeze is the opposite. When the ocean is warmer than the land, air movement changes from land to water. When this happens in the summer, more electricity is used to run air conditioners and room fans.

Wind speed is measured using an anemometer. The anemometer has a propeller that turns in the wind. The rate at which the propeller turns can be converted to a measurement of speed. Wind is measured in miles per hour (mph), kilometers per hour (kph), or in knots. Knots are a measurement used by airplane pilots and sailors.

Wind Scale	
0-12 mph	light
13-24 mph	moderate
26-31 mph	strong
32-63 mph	gale
64-73 mph	storm
More than 73 mph	hurricane

AIR PRESSURE

THE BIG IDEA
When pressure is high, Look for blue sky. When pressure is low, Expect rain or snow.

Air pressure is measured using a barometer. High pressure means fair weather. Low pressure means a change in weather, usually to foul weather.

CLOUDS

Rising warm air can hold more water than cold air. The water will rise with the air molecules. As the air cools, the water condenses to form clouds. The table below shows the common types of clouds and the weather conditions associated with them. Clouds form at the boundary between cold and warm air fronts.

Cloud Type	Cloud Name	Type of Weather
	Cumulonimbus	Thunderstorms, tornadoes, hurricanes
	Cumulus	Fair weather
	Cirrus	Light rain or snow could arrive in a few days
	Nimbostratus	Rain, snow, drizzle
	Stratus	Weather may be changing

HUMIDITY

"It's not the heat. It's the humidity." Water in the air can make the air temperature feel warmer than it is.

Moisture in the air is called **humidity**. Relative humidity is a measurement of how much water is in the air compared to how much it could hold. When the air is at 100 percent humidity, the air can hold no more water. Relative humidity depends on air temperature because warm air can hold more water than cold air.

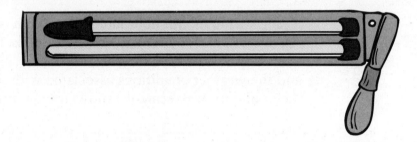

A **psychrometer** is a tool used to measure relative humidity. The psychrometer has two thermometers. One thermometer has a small wick of cotton covering the bulb. Drops of water are added to the cotton wick. This thermometer is called the **wet bulb**. The other thermometer has no covering over the bulb. This thermometer is called the **dry bulb**.

The two thermometers are usually in a case attached to a handle that can rotate. The psychrometer is swung around by the handle for a few minutes. The temperature of each thermometer is then read. The difference between the wet-bulb temperature and the dry-bulb temperature can be used to calculate the relative humidity. Reference charts are used to quickly find the relative humidity.

When the psychrometer is swung through the air, water evaporates from the cotton on the wet-bulb thermometer. Evaporation causes a cooling effect. If the air is dry, the reading from the wet-bulb thermometer will be much lower than the dry-bulb thermometer. When the air is humid, the difference between the two thermometers will be less. When both thermometers record the same temperature, the relative humidity is 100 percent. Dew, fog, and precipitation can form when the humidity reaches 100 percent.

During the day, water evaporates into the air as the Sun warms the surface of the Earth. At night, temperatures begin to cool. Water may condense onto surfaces. When this happens, air is said to be at the **dew point**. The next morning dew may still be on the grass. The cycle begins again.

Fog is surface level condensation in the air. Water drops out of the air as the air makes contact with the cooler ground. You may have noticed this over a soccer field in the spring or fall when the nights are cooler.

PRECIPITATION

The type of precipitation that falls from clouds depends on air temperature. The table below shows conditions needed for common types of precipitation.

Precipitation Type	Conditions
Snow	Water droplets condense around a dust particle and freeze when air temperatures are below 32 degrees Fahrenheit.
Hail	Water begins to freeze and either is thrown up and down in cumulonimbus clouds or falls very slowly so that it adds ice layers around the original ice particle.
Sleet	Water falls as rain and freezes near ground level into tiny pellets of ice.
Freezing Rain	Water falls as rain and freezes on contact with ground surfaces.
Rain	Water in liquid form falls in droplets to the ground.
Drizzle	Water in liquid form falls in a fine mist or scattered droplets.

BIG IDEA CHECK-UP 6.5 (For answers, see page 164.)

Multiple Choice (Circle the correct answer.)

1. Which tool would be used to measure relative humidity?

   A. Anemometer   C. Psychrometer

   B. Barometer   D. Thermometer

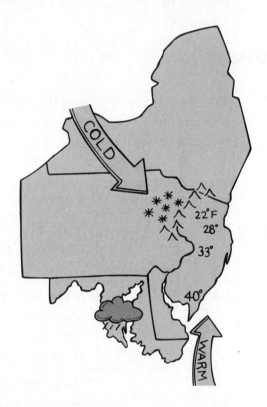

2. The weather map above appeared in the local morning newspaper. Which of the following would be the weather forecast for New Jersey?

   A. Expect rain in southern New Jersey, a mix of rain and snow in central New Jersey, snow in northern New Jersey.

   B. Expect rain to begin in southern New Jersey before spreading north and west with heaviest rainfall in the Newark Bay area.

   C. Expect snow in northern New Jersey to spread east and south with snowfall totals highest in the Cape May area.

   D. Expect freezing rain in southern New Jersey, sleet in central New Jersey, and heavy snowfall in northern New Jersey.

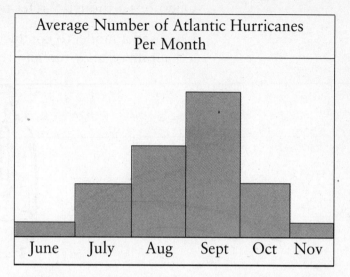

Month	June	July	Aug	Sept	Oct	Nov
Average Ocean Temperature at Atlantic City, NJ in degrees Fahrenheit	62–64	69–70	72–73	72–67	63–58	53

Use the graph and data table to answer question 3.

3. Hurricanes need warm ocean water. When would New Jersey be most likely to experience the effects of a hurricane?

A. June–July

B. July–August

C. August–September

D. September–October

4. Which hurricane danger is greater for New Jersey shore communities than for inland communities?

A. Storm surge

B. Flooding

C. Heavy rains

D. Landslides

Questions 5 though 9 refer to the experiment described below.

Levi used a model to conduct an experiment. In his journal he drew a diagram of his setup and recorded his observations.

Levi used two conditions. He compared the results of one setup with a lamp turned on for 5 minutes and then turned off for 10 minutes to an identical setup without a lamp.

Data and Observations

Setup with Lamp	Air Temperature (degrees Fahrenheit)	Observation
Lamp turned off	72	Water was in the bottom of the bottle. The sides of the bottle were dry.
Lamp turned on for 5 minutes	90	The thermometer began to swing around. The temperature increased. Small water drops started to form near the top of the bottle.
Lamp turned off for 10 minutes	78	The thermometer continued to twist around. More water collected at the top of the bottle. Water began to run down the sides of the bottle.

Setup without Lamp	Air Temperature (degrees Fahrenheit)	Observation
Start	72	Water was in the bottom of the bottle. The sides of the bottle were dry.
After 15 minutes	72	Water was in the bottom of the bottle. The sides of the bottle were dry.

5. Which part of the experiment was the lamp?

    A. Hypothesis

    B. Control

    C. Variable

    D. Question

6. The thermometer moved in the bottle with the lamp. Which statement does this support?

    A. Hot air rises and cold air sinks.

    B. Warm air can hold more water.

    C. Hot air causes evaporation.

    D. Cold air causes condensation.

7. Which part of the bottle with the lamp would experience the lowest pressure?

    A. The top when the lamp was turned off

    B. The sides when the lamp was turned off

    C. The entire bottle when the lamp was turned on

    D. The bottom when the lamp was turned on

8. If the bottle were empty, which of the following would **not** change when the lamp was turned on?

    A. Relative humidity

    B. Air pressure

    C. Wind speed

    D. Temperature

9. In order to observe cloud formation, which of the following would Levi need to add to the experiment?

A. More heat

B. Ice

C. Dust particles

D. More water

Check your answers before going on to the next topic.

6.6 PROCESSES THAT SHAPE THE EARTH

6.6.A VOLCANOES

> **THE BIG IDEA**
>
> Volcanoes can cause rapid changes in Earth's surface.

Volcanoes rapidly change the shape of the land. A red-hot lava flow destroys all life in its path. As the lava cools, caves may form.

Over time, lichens and mosses will grow on the rocky surfaces. With more time, biological weathering will begin to build soil for grasses and other plants with root systems.

Scientists study the processes of destruction and renewal on the slopes of Hawaii's volcanoes and the volcanoes of the northwestern United States. Volcanoes may be quiet or dormant for many years and then awake with great force. In 1980, Mount St. Helens erupted blowing out a large area of rock from the side of the mountain.

Dinosaurs and reptiles were witnesses to volcanic activity in New Jersey. When the Atlantic Ocean began to form around 200 million years ago, magma deep in the Earth hardened to become the diabase rocks of the Palisades in northeastern New Jersey. Basalt rock that was the ancient sea floor is a ridge that runs through modern-day Bergen County and Passaic County. The towering basalt walls of the Great Falls of Paterson, New Jersey, are evidence of the violent birth of the Atlantic Ocean. More evidence can be seen in the turtle back rocks of Morristown and the Watchung Mountains.

6.6.B EARTHQUAKES

THE BIG IDEA

Earthquakes can cause sudden changes in Earth's surface.

Earthquakes can change the landscape by raising land and tearing deep trenches as the ground moves apart. Sometimes land lifted by one earthquake will drop during the next. In sandy, wet soil, earthquakes can cause flooding because the quake vibrations shake water out of the sand.

Earthquakes send out two types of waves. The first wave moves quickly in a straight line over great distances. The second wave shakes the surface back and forth. The second type of wave causes the most damage.

Earthquakes occur along fault lines in Earth's crust. California is well known for the San Andreas Fault and for earthquakes. Earthquakes are common around the Pacific Ocean rim because of volcanoes. This area is called the Ring of Fire.

Earth's plates can rub past each other at the edges. Sometimes the forces holding the two giant rock sheets fail, and the plates slip. This causes vibrations in the ground and shaking at the surface.

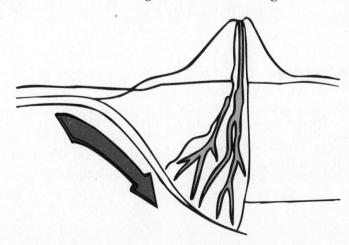

In other locations, a heavier plate will slide under a lighter plate. This usually happens where ocean plates meet continental (land) plates. The rock being pushed deep into the earth may melt and become magma in a chamber underneath a volcano. When pressure builds up—boom!—the volcano erupts and the ground shakes. (See figure on bottom of page 121.)

A third type of earthquake happens when plates of equal density move toward each other and collide. This type of collision builds up mountains such as the Himalayas. Earthquakes along continental collisions are often violent and cause great damage and loss of life.

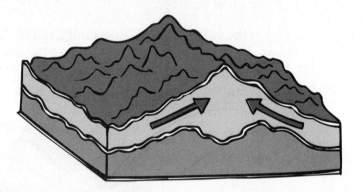

New Jersey has fault lines. Little tremors (vibrations) will cause minor earthquakes along the Ramapo Fault in northern New Jersey. Since the late 1600s, 400 earthquakes have been reported from Salem County to Bergen County. Cheesequake is one of New Jersey's more active earthquake sites.

The mountains of the New Jersey highlands may once have been as tall as the Rockies or even the Himalayas. They were formed in much the same way 1.3 billion years ago by rock lifting rock upward into great mountain ranges. The great pressure also created the metamorphic rocks gneiss and marble.

6.6.C EROSION

THE BIG IDEA
Erosion changes Earth's surface as the result of the movement of water.

Water—soft, gentle, and fluid—will wear away hard rock, carving great natural wonders such as the Grand Canyon. Erosion is one process that turns rocks into sediments.

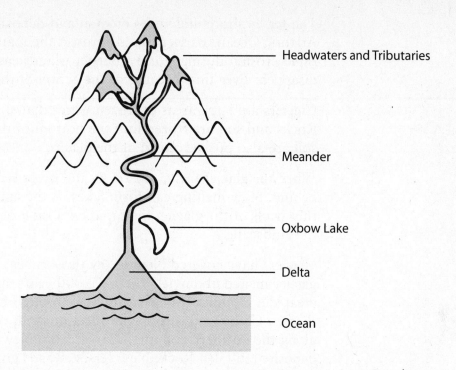

Headwaters and Tributaries

Meander

Oxbow Lake

Delta

Ocean

Rain and melted snow run off the rocky surfaces of mountains. Over time, water will cut a channel that becomes a mountain stream. Springs and streams are called the headwaters of a river. As streams come together they become tributaries (contributors) to larger rivers. Rivers follow the terrain and collect additional water from rainwater runoff.

Sometimes rivers curve away from a straight path. The water moves faster on the outside of the bend eroding sediments from the bank. The sediments are deposited on the inside of the bend. This creates a meander. An ox bow lake may form if a meander becomes cut off from the main branch of the river by erosion and deposition.

Rivers eventually drain to an ocean. As the river nears the ocean, the land becomes flat. The sediments carried downstream pile up at the mouth of the river. This forms a delta. The best example of a delta in the United States is the Mississippi River Delta around New Orleans. New Jersey's rivers often end in a marsh before flowing into a bay.

Raised landforms create a watershed. Water flows from a higher elevation to a lower elevation. New Jersey's rivers flow into Newark Bay, the New York Harbor Bight, Atlantic Ocean, Delaware River, or Delaware Bay depending on where they begin.

Running water moves soil around. A flash flood can carry away topsoil destroying valuable cropland. Flash floods can also cause landslides.

The Jersey shore undergoes erosion and deposition by action of the Atlantic Ocean. Barrier islands protect the main shoreline from rapid erosion during storms. Barrier islands can appear and disappear over time because of the action of the ocean waves.

Glaciers are large areas of frozen water that slowly move downhill. Rocks and soil are frozen into the bottom of the glacier. Rocks and soil are also pushed ahead of the glacier.

When the glacier melts, the rocks and other sediments are left behind. Slow melting can form lakes. If the lakes burst through the thin walls of the glacier, the rapid explosion of water will reshape the landscape.

Glaciers have covered New Jersey three times. The most recent glacier melted about 10,000 to 12,000 years ago. The glacier created a mound of rocks at its edge through Union, Morris, and Warren Counties. It dropped 80 feet of sand, rocks, and pebbles along the northern counties. A rock from New England was deposited in Glen Rock, New Jersey. Roads go around the rock because it was too large to move.

The melting waters carved out the Passaic and Hackensack Rivers. Scrape marks on rocks in Jenny Jump State Park show the abrasive action of the bottom of the glacier.

6.6.D WEATHERING

THE BIG IDEA
Weathering slowly causes changes in Earth's surface.

Weathering breaks up rocks. Water from precipitation will run into small cracks in rocks. As the temperature drops, the water freezes, expands, and widens the crack. If this process happens over and over, the rock may break apart. This same process causes manmade materials such as sidewalk concrete and street blacktop to crack and develop potholes.

Plant seeds sometimes fall into the widened cracks. The roots of the plant will further break up the rock. To see the power of plants, place a bean seed in plaster of Paris. Add water. Right before your eyes, the bean will sprout and force its way through the hardened plaster.

Small animals will dig into cracks in the rock causing rocks to break or fall off ledges. Action by plants and animals is called biological weathering.

Another type of weathering is chemical weathering. Acids in rain react with chemicals in rocks causing the rocks to dissolve. You can observe this by placing a piece of chalk in vinegar. Watch how long it takes for the chalk to disappear.

BIG IDEA CHECK-UP 6.6 (For answers, see page 165.)

Multiple Choice (Circle the correct answer.)

1. Which of the following would cause a slow change in a landscape?

    A. An earthquake

    B. A volcanic eruption

    C. A landslide

    D. Weathering

2. Which of the following created the Watchung Mountains of New Jersey?

    A. Erosion and deposition

    B. Volcanic eruptions

    C. Earthquake uplift

    D. Glaciers

3. A rock containing 14,000-year-old fossils from upstate New York was found in the Newark, New Jersey, area. Which best explains how it ended up in the Hackensack Meadowlands?

    A. It was deposited by the Hudson River.

    B. It was deposited by erosion at the headwaters of the Hackensack River.

    C. It was deposited by a glacier during the last ice age.

    D. It was deposited by an earthquake along the Ramapo Fault.

4. Still Run and Scotland Run flow out of the Pine Barrens. They come together to form the Maurice River. Which term describes Still Run and Scotland Run?

   A. Mouth

   B. Ox bow lake

   C. Meander

   D. Headwaters

5. Which would be more likely to experience erosion?

   A. Igneous rock from magma

   B. Igneous rock from lava

   C. Sedimentary rock

   D. Metamorphic rock

6. In the spring, a hiker in Wawayanda State Park in northern New Jersey noticed a pile of large flat chunks of rock at the base of a cliff made of the same type of rock. Which statement best explains why the rocks broke away from the cliff?

   A. Earthquakes along the Ramapo Fault shook the rocks loose.

   B. Freezing and thawing caused the rocks to crack and fall.

   C. A moving glacier broke the rocks by abrasion.

   D. Erosion by a flash flood caused the rocks to break up.

Check your answers before going on to the next topic.

6.7 HOW WE STUDY THE EARTH

THE BIG IDEA
Maps can be used to show the physical features of Earth.

Maps can be used to show earthquake activity, watersheds, elevation, the location of rock types, and weather patterns.

Before computers, maps were drawn by hand. Cartographers are people who draw maps. To show multiple types of information, the cartographer may have begun by drawing a basic outline map. Then the cartographer would create transparent overlays that would build up layers of information onto the base map.

Computers construct maps from data. The data come from many sources. Scientists working in the field may collect data directly. Earthquake monitors or weather stations send data automatically to a computer. Satellites may collect data and take pictures from space.

Geographic information systems use data from many sources to create detailed maps. These maps are often used for planning purposes.

Wind density maps were used to plan the location of wind turbines in Atlantic City, New Jersey. Maps of bird migratory patterns based on radar data were used in planning the location of New Jersey's first off-shore wind farm.

BIG IDEA CHECK-UP 6.7 (For answers, see page 165.)

Multiple Choice (Circle the correct answer.)

1. Which of the following would provide the most accurate map of the path of a hurricane in the North Atlantic?

    A. Radio reports from ships at sea.

    B. Observations from airline pilots.

    C. Weather stations on islands.

    D. Satellite images of the North Atlantic.

2. Which map would be important for planning a new community in southern California?

    A. A map of global earthquake ativity

    B. A map of local fault lines

    C. A map of west coast plate boundaries

    D. A map of California fault lines

Check your answers before going on to the next topic.

ASTRONOMY AND SPACE SCIENCE

ASTRONOMY AND SPACE SCIENCE CONCEPT MAP

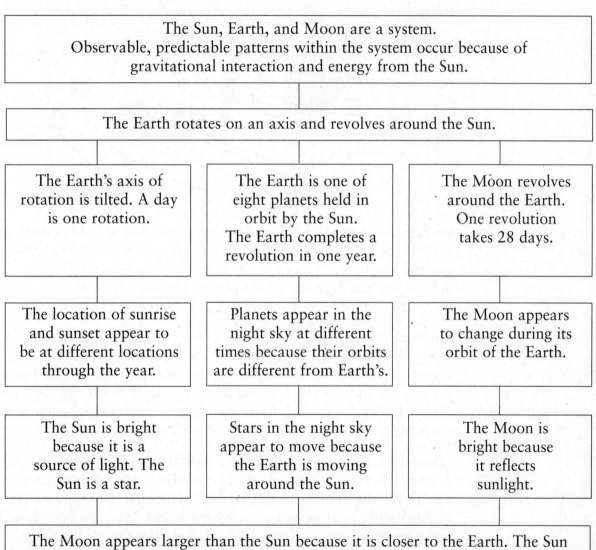

The Sun, Earth, and Moon are a system.
Observable, predictable patterns within the system occur because of gravitational interaction and energy from the Sun.

The Earth rotates on an axis and revolves around the Sun.

The Earth's axis of rotation is tilted. A day is one rotation.

The Earth is one of eight planets held in orbit by the Sun.
The Earth completes a revolution in one year.

The Moon revolves around the Earth.
One revolution takes 28 days.

The location of sunrise and sunset appear to be at different locations through the year.

Planets appear in the night sky at different times because their orbits are different from Earth's.

The Moon appears to change during its orbit of the Earth.

The Sun is bright because it is a source of light. The Sun is a star.

Stars in the night sky appear to move because the Earth is moving around the Sun.

The Moon is bright because it reflects sunlight.

The Moon appears larger than the Sun because it is closer to the Earth. The Sun appears larger than other stars because it is the closest star to Earth. Objects that are closer appear to be larger than objects that are far away.

7.1 EARTH, MOON, SUN SYSTEMS

THE BIG IDEA

Planets orbit the Sun. Moons orbit planets. Planets are held in orbit by the gravity of the Sun. Moons are held in orbit by the gravity of a planet.

The Sun is a massive ball of hydrogen, helium, oxygen, and other elements including carbon and iron. The inside of the Sun could hold more than 1.3 million Earths. If the Sun were drawn as a circle, it would be 109 Earth's wide. The mass of the Sun is equal to the mass of 332,950 Earths. That's huge!

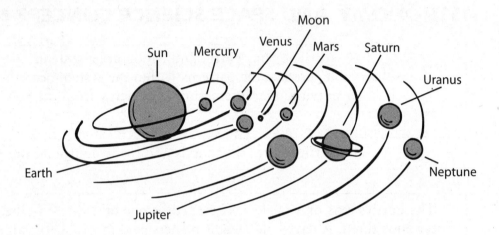

Because the mass of the Sun is so great, it is able to hold planets such as the Earth in orbit. The planets are held in orbit by the Sun's gravity. The time it takes for a planet to complete a revolution around the Sun depends on the planet's distance from the Sun. One complete revolution is a year. For Earth that is 365 days.

Think of the planets as runners in different track lanes. The outside lane of the track is a longer distance than the inside lane. If all the planets were in a line at the start of the race, Mercury would finish the race around the Sun first and Neptune would finish last. Mercury circles the Sun in about 3 Earth months. Neptune takes nearly 165 Earth years to complete one orbit of the Sun.

Planets also have gravity. Smaller rock-like objects often orbit planets. The Moon is Earth's natural satellite. The table on page 131 shows the number of moons each planet has. The data are as of 2005 when new moons were discovered by space probes. New space missions may discover additional moons around the outer planets.

Planet	Number of Moons (as of 2005)
Mercury	0
Venus	0
Earth	1
Mars	2
Jupiter	63
Saturn	60
Uranus	27
Neptune	13

The solar system has dwarf planets that orbit the Sun, but do not have enough gravity to clear their neighborhood of similar-sized objects. Pluto was reclassified from planet to dwarf planet in 2006.

Sometimes when a planet is passing Earth in its orbit or being passed by Earth, it will appear in Earth's night sky. They are bright objects that do not flicker like stars. They reflect sunlight and are closer to Earth than stars.

Other objects are also orbiting the Sun. Comets are ice and dust that have predictable paths around the Sun. Comets appear to move across the sky and have glowing tails.

Large chunks of rock called asteroids are found in orbit between Jupiter and Mars. Sometimes one of the smaller rocks will break loose and become a meteoroid moving through space. If a meteoroid enters Earth's atmosphere, it becomes a meteor. It the meteor fails to burn up from the friction of Earth's atmosphere, it will hit the ground as a meteorite. Burning meteors are called falling stars even though they are not stars.

April, August, and November are good months to view meteor showers in New Jersey. Most meteorites cause little damage, but large ones may have changed Earth's climate in the past. One large impact site formed what is now Chesapeake Bay. An impact 65 million years ago off the Yucatan Peninsula of Mexico may have ended the age of the dinosaurs.

BIG IDEA CHECK-UP 7.1 (For answers, see page 166.)

Multiple Choice (Circle the correct answer.)

1. Which of the planets listed below would have the longest orbit around the Sun?

   A. Mars

   B. Jupiter

   C. Saturn

   D. Uranus

2. Which of the following pairs of space objects share the same classification?

   A. Meteoroid and Sun

   B. Venus and Mars

   C. Earth and Moon

   D. Asteroids and Comets

Check your answers before going on to the next topic.

7.2 ROTATION

THE BIG IDEA
Earth rotates on an angle around its axis in a 24-hour period called a day.

Rotation is movement in a circle around a center point. Think of a figure skater in a spin. One foot stays in place while the skater's body moves in a circle around a point near the toe pick of the skate blade.

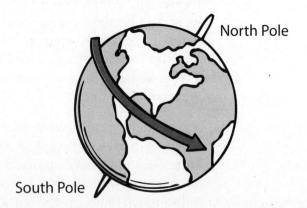

When an object rotates around a center point inside its own body it is said to spin around an axis. In the above diagram, the earth's axis is shown as a line passing from the North Pole through the center of the Earth to the South Pole. The Earth's axis is tilted.

It takes the Earth 24 hours to complete one rotation around its axis. The time it takes for a planet to complete one rotation around its axis is called a day. Because the Earth is turning, the Sun appears to rise on the eastern horizon and set on the western horizon.

BIG IDEA CHECK-UP 7.2 (For answer, see page 166.)

Multiple Choice (Circle the correct answer.)

1. Which of the following diagrams correctly shows how the Sun lights the Earth?

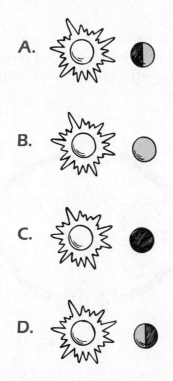

A.

B.

C.

D.

Check your answer before going on to the next topic.

7.3 REVOLUTION

THE BIG IDEA
The Earth revolves around the Sun while rotating on a tilted axis.

In New Jersey, the Sun appears to change location at sunrise during the year. This is something noticed by ancient people from many cultures and locations around the world.

Archeologists, scientists who study ancient humans, observed that villages constructed by mound builders 2,000 years ago in the Ohio Valley were laid out in a circular pattern. They conducted an experiment and discovered that the main structures were built so that sunlight would cast shadows toward specific buildings on the first day of spring, the first day of summer, the first day of fall, and the first day of winter. The buildings found at Fort Ancient were a solar calendar.

As the Earth rotates on its tilted axis, it moves around the Sun. Movement around another body is called **revolution**. It takes the Earth 365 days to complete one revolution around the Sun, or one orbital period.

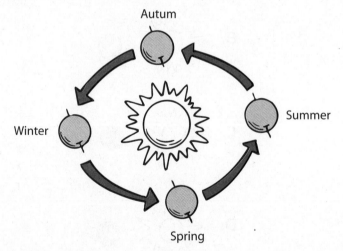

Earth's axis is tilted. Because of this tilt, different places on Earth receive more or less energy from the Sun. The equator and tropical regions receive the most energy. The poles have the greatest variation in solar energy. Areas between the tropics and the poles have four seasons. New Jersey is located in this region.

Hours of Sunlight in New Jersey

Date	March 21	June 21	September 23	December 21
Daylight	12 hours	15 hours	12 hours	9 hours

Throughout the year, the time between sunrise and sunset changes in New Jersey because the Earth revolves around the Sun and rotates on a tilted axis. The longest days are in the summer when the axis tilts toward the Sun in the northern hemisphere. The shortest daytime period in New Jersey is the first day of winter when the Earth tilts away from the Sun. On the first day of spring and the first day of summer, all places on Earth receive the same hours of sunlight.

June 21	March 21 September 23	December 21
Southeast	East	Northeast

Although the Sun rises in the east and sets in the west, the tilt of the Earth on its axis explains why the sunrise seems to change locations through the year. This also explains why shadows are longer in the winter and shorter in the summer at noon.

BIG IDEAS CHECK-UP 7.3 (For answers, see page 166.)

Multiple Choice (Circle the correct answer.)

1. Which date has the least number of hours between sunrise and sunset?

   A. The first day of spring

   B. The first day of summer

   C. The first day of fall

   D. The first day of winter

2. Students in New Jersey measured the shadow cast by a meter stick at noon on the first day of fall, the first day of winter, the first day of spring, and the first day of summer.

Fall Students measured a medium length shadow
Winter Students measured a longer shadow
Spring Students measured a medium length shadow
Summer Students measured a short shadow

Which of the following did the class prove?

A. The Earth is tilted on its axis.

B. The Earth orbits the Sun.

C. The Earth has a 24-hour day.

D. The Earth is closer to the sun in the winter.

Check your answers before going on to the next topic.

7.4 MOON PHASES

THE BIG IDEA

The moon revolves around the Earth and reflects sunlight from its surface. The changes in the moon's appearance are called phases.

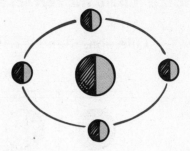

The moon revolves around the Earth. It takes approximately 28 days for the moon to complete one revolution. The moon's appearance changes during the 28-day orbit. The changes are called phases of the moon.

Many cultures have measured time using the moon phases. Lunar calendars have months that are measured from one new moon to the next. Lunar years are different than solar years because they are not based on the revolution of the Earth around the Sun.

The moon is rock. It does not produce light. It appears to be bright because it reflects sunlight. The amount of light reflected depends on how much of the moon's surface is reflecting sunlight. Black on the diagram indicates that there is no reflection. The four main phases of the moon, as they appear to someone on Earth, are shown below.

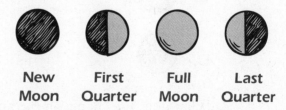

| New Moon | First Quarter | Full Moon | Last Quarter |

If the moon is gray rock why does it appear to be white? If a lit flashlight is pointed at a black piece of paper in a darkened room, the paper will appear white. It is the same optical illusion that explains why a moon of dark gray rock can appear white.

BIG IDEA CHECK-UP 7.4 (For answers, see page 166).

Multiple Choice (Circle the correct answer.)

Questions 1 and 2 refer to the diagram below.

Kayla's Moon Phase Journal

April 4	April 11	April 14	April 18	April 26	May 3

1. Which of the following diagrams would Kayla have drawn in the space labeled April 14?

A.

B.

C.

D.

2. What does the change in the appearance of the Moon prove?

   A. The Moon completes one rotation on its axis every 28 days.

   B. The Moon completes one orbit of the Earth in 28 days.

   C. The Moon produces different amounts of light during 28 days.

   D. The Moon disappears into the Earth's shadow every 28 days.

Check your answers before going on to the next topic.

7.5 STARS, GALAXIES, AND THE UNIVERSE

THE BIG IDEA

The Sun is a star that belongs to the Milky Way galaxy, one of many galaxies in the universe. Stars have different characteristics.

The Sun is Earth's closest star. Compared to other stars the Sun is a medium-sized yellow star. Our solar system is one among millions in the Milky Way galaxy. Our galaxy is one of many that make up the universe.

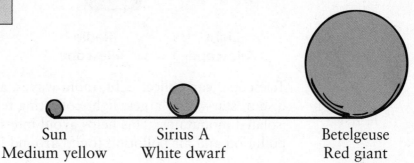

Sun	Sirius A	Betelgeuse
Medium yellow	White dwarf	Red giant

The table shows the relative sizes of three types of stars. Betelgeuse is a red giant star in the Orion constellation and can be seen in New Jersey during the winter months. Sirius A is a white dwarf star that is the brightest star in the constellation Canis Major, visible in the winter in New Jersey. After the Sun, Sirius is the nearest star to Earth.

The colors of stars depend on the elements in the stars, the heat of the stars, and the distance from Earth. Very hot stars may appear closer than cooler stars even though they are much farther from Earth. Very small stars that are near Earth may appear brighter than larger stars that are farther away.

Constellations are patterns of stars that humans see in the sky. Some stars of the same constellation may be very far from each other. Constellations were used for navigation long before the invention of satellite-based global positioning systems.

The North Star is Earth's current fixed star. The Earth's axis always tilts toward the North Star. Therefore, the North Star does not appear to move across the night sky, as do other stars.

Astronomers, scientists who study the stars, use land-based telescopes, an orbiting telescope, and space probes to gather information about our galaxy and other galaxies in the universe.

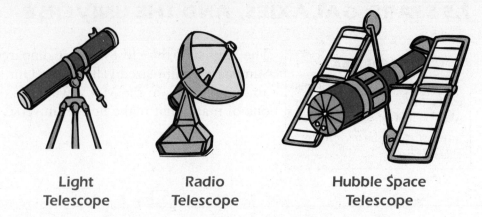

Light Telescope   **Radio Telescope**   **Hubble Space Telescope**

Telescopes can collect light, radio waves, and other energy from distant stars. The largest light-collecting telescopes are located on isolated mountains. This helps avoid interference from light pollution and air pollution found around large cites.

The Hubble space telescope has allowed astronomers to see objects in space at greater distances and with greater clarity than land-based telescopes. The Hubble space telescope has recorded images of nebula, exploding stars, and never before seen parts of the universe.

BIG IDEA CHECK-UP 7.5 (For answers, see page 167.)

Multiple Choice (Circle the correct answer.)

1. Which of the following is **not** located in the solar system?
   A. Planet
   B. Comet
   C. Constellation
   D. Moon

2. What star can only be seen during the day from Earth?
   A. Sirius A
   B. Betelgeuse
   C. Sun
   D. North Star

Check your answers before going on to the next topic.

ENVIRONMENTAL SCIENCE

ENVIRONMENTAL SCIENCE CONCEPT MAP

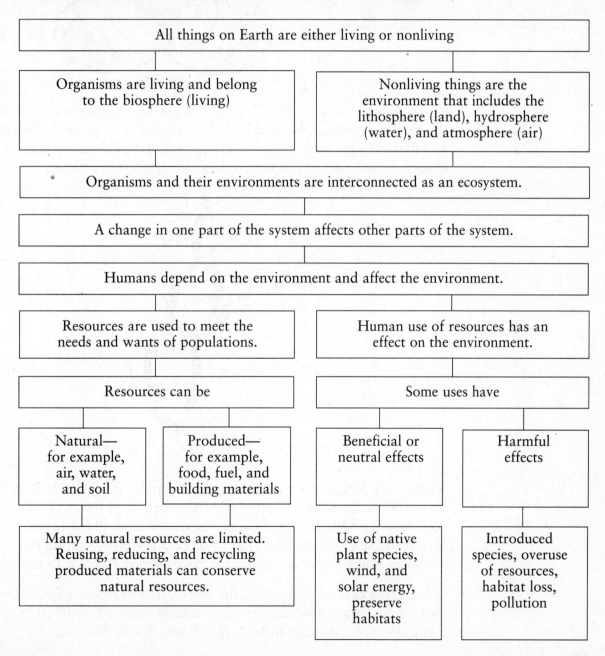

All things on Earth are either living or nonliving

Organisms are living and belong to the biosphere (living)

Nonliving things are the environment that includes the lithosphere (land), hydrosphere (water), and atmosphere (air)

Organisms and their environments are interconnected as an ecosystem.

A change in one part of the system affects other parts of the system.

Humans depend on the environment and affect the environment.

Resources are used to meet the needs and wants of populations.

Human use of resources has an effect on the environment.

Resources can be

Some uses have

Natural— for example, air, water, and soil

Produced— for example, food, fuel, and building materials

Beneficial or neutral effects

Harmful effects

Many natural resources are limited. Reusing, reducing, and recycling produced materials can conserve natural resources.

Use of native plant species, wind, and solar energy, preserve habitats

Introduced species, overuse of resources, habitat loss, pollution

8.1 ECOSYSTEMS

THE BIG IDEA

Living things need shelter, food, water, and air for survival. Resources found in their surroundings—the environment—meet these needs. The interactions of living things with each other and with nonliving things form an ecosystem.

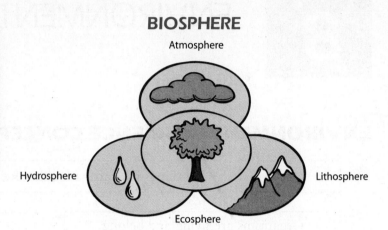

BIOSPHERE

The health of an ecosystem depends on the use of resources within the local ecosystem, regional biome, or global environment. Energy and materials cycle through the ecosystem in a delicate balance. A change in one part of the system has effects on other parts.

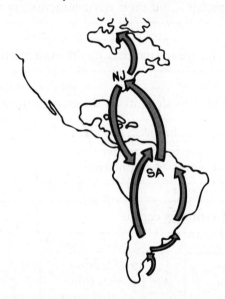

New Jersey forests provide habitat for animals such as migratory birds. When forests are cleared for construction of human communities, wild habitat is lost. Migratory birds breed in the forests of New Jersey. When the birds return from spending the winter in South America, they have fewer places to build nests.

When birds cannot reproduce, the population decreases. Fewer birds fly back to South America for the winter. A reduced population of birds affects ecosystems in both North and South America.

New Jersey forests protect water supplies. Trees hold soil and filter rainwater. Without trees loose soil and pollution will run in streams and rivers. The sediments and pollutants may kill insect larvae that need clear, clean water. Fish and other animals that depend on insects or insect larvae will have less food. The plants and animals of the stream ecosystem will change.

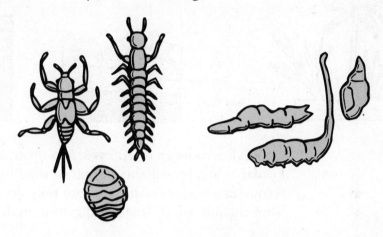

Clean water insects   Polluted water insects

Water for human use will also be affected. Water from sources protected by forests is less expensive to process for human use. Water from other sources may require more expensive processing steps to provide clean drinking water.

8.2 RENEWABLE VS. NONRENEWABLE RESOURCES

THE BIG IDEA
Humans use resources from the environment for fuel, food, clothing, and other needs and wants. Natural resources can be renewable or nonrenewable.

Renewable resources can be replaced in less than 100 years. For example, when trees are cut down to be turned into paper or houses, new trees can be planted. Trees, water, and soil are considered renewable resources as long as the use is balanced with replacement.

Nonrenewable resources cannot be replaced in less than 100 years. Metals and some minerals are nonrenewable. Oil and coal from ancient sources are nonrenewable. They are called fossil fuels since they came from plants and animals that died millions of years ago.

For example, most of the coal used today formed over 300 million years ago. When it is burned as fuel, it is gone. The environment that changed dead plants to coal no longer exists.

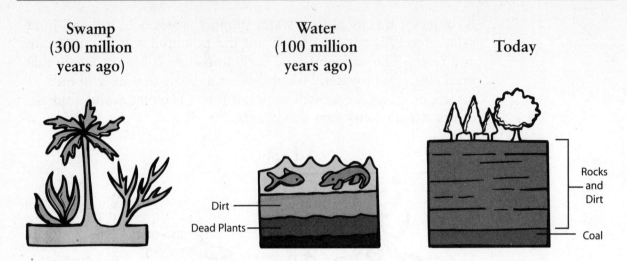

Swamp
(300 million
years ago)

Water
(100 million
years ago)

Today

Dirt

Dead Plants

Rocks
and
Dirt

Coal

Coal takes more than 100 years to form. Future coal will not be available in a reasonable period of time for use. Some energy resources are sustainable because they do not need to be replaced. They include wind, solar, and geothermal.

8.3 HUMAN IMPACTS ON THE ENVIRONMENT

THE BIG IDEA
Human activities affect water, land, and air.

Litter, toxic wastes, and smog pollute the environment affecting all living things. Sustainable resource management is planned use of resources to meet the needs of humans today while planning for the needs of humans in the future.

Coal can be used to produce electricity for human use. Coal is a nonrenewable resource. When coal is burned, air pollution increases. Greenhouse gases also increase.

Greenhouse gases trap heat in the atmosphere. Increasing heat may cause glaciers to melt. Melting glaciers add water to the oceans. Rising sea levels can cause flooding and cause salt water to move farther into tidal rivers.

The Delaware River is a tidal river. Water flows from the Delaware River into Delaware Bay when the tide is going out. When the tide comes in, salt water flows up the Delaware River.

The place where salt water meets fresh water is called the salt line. Salt water in the river affects freshwater in southern New Jersey wells. The salt line of the Delaware River has moved north. This may be a result of rising sea levels.

Electricity can also be generated using energy from the Sun or wind. These sources are examples of sustainable energy production. Solar energy is already being used by homeowners, businesses, and schools in New Jersey.

Atlantic City is using wind power to produce electricity. A planned wind farm will provide enough electricity for 200,000 homes in New Jersey.

Solar and wind power do not contribute greenhouse gases to the atmosphere and do not produce harmful air pollutants.

Meeting the needs of humans today while considering the needs of future humans requires careful planning and difficult decision making. Choices have pluses and minuses.

New Jersey has plenty of wind off the Jersey shore. Placing a wind farm in the ocean will provide a sustainable source of electricity for New Jersey. However, the same conditions that make this a great location for a wind farm make it a popular route for migratory birds. The effect of placing a wind farm in this important flyway is unknown. Very careful planning includes studies of bird migration patterns to avoid an ecological disaster.

Conservation saves resources. Recycling, reductions of use, and reuse of materials are conservation practices.

Recycle

Plastic bottles and metal cans end up in a landfill as trash. Landfills produce toxic substances that may contaminate drinking water or soil used for food crops.

Plastic bottles and metal cans may be recycled into other products. Recycling reduces the use of nonrenewable resources such as oil and ores.

Reducing the use of plastic and metal cans also reduces the depletion of nonrenewable resources. Finding ways to package food and beverages in reusable containers would be a conservation practice.

Conservation includes simple everyday decisions by individuals. Turning off lights as you leave the room, recycling paper, and composting food scraps help save Earth's resources.

BIG IDEA CHECK-UP CHAPTER 8 (For answers, see page 167.)

Multiple Choice (Circle the correct answer.)

1. A renewable resource is something that can be replaced in less than 100 years. Which of the following comes from a renewable resource?

   A. Gasoline from oil

   B. Wood from trees

   C. Natural gas from coal

   D. Aluminum from ore

2. Conservation saves resources. Which of the following is an example of conservation?

   A. Letting the water run when washing paintbrushes

   B. Drawing with crayons made from soybeans

   C. Leaving lights on in an empty classroom

   D. Using metal staples to hold paper together

3. Sustainable energy means that energy source will continue to be available. Which of the following is not sustainable?

   A. Electricity from solar energy

   B. Electricity from wind energy

   C. Electricity from tidal energy

   D. Electricity from coal energy

Open-Ended Question

4. Mr. Yukawa's class divided the lunchroom trash into bags of trash that could be recycled, composted, or reduced by replacement with something reusable. The bags were labeled "plastic bottles," "aluminum cans," "juice boxes," "food waste", and "other." The "other" bag contained waste such as yogurt cups, sandwich bags, plastic spoons and forks, soiled napkins, and other items.

Bags of Lunchroom Trash Collected for 5 Days

Number of Bags				
5			5	5
	4	4		
Plastic Bottles	Aluminum Cans	Juice Boxes	Food Waste	Other Waste

The chart shows the amount of trash by category collected during 5 days of school.

Give two examples of things that students could do to reduce the lunchroom trash and conserve resources. Explain why your suggestion will conserve resources.

1. _____

_____

_____

2. _____

_____

_____

ANSWER KEY TO PRACTICE QUESTIONS, CHAPTERS 1-8

CHAPTER 1 TOOLS OF THE SCIENTIST

CHECK YOUR UNDERSTANDING 1.1

The cylinder is balanced by a known mass of 10 grams; therefore the mass is 10 grams.

CHECK YOUR UNDERSTANDING 1.2

The volume of the box is 100 cm^3.

First calculation: 10 cm × 5 cm = 50 cm^2

Second calculation: 50 cm^2 × 2 cm = 100 cm^3

CHECK YOUR UNDERSTANDING 1.3

There are 40 ml of water in the cylinder because that is the lowest part of the curve (meniscus).

CHECK YOUR UNDERSTANDING 1.4

The density is 2 grams per cubic centimeters or 2 g/cm^3.

CHECK YOUR UNDERSTANDING 1.5

The distance is 25 millimeters.

CHECK YOUR UNDERSTANDING 1.6

A good tool would be a calendar.

CHECK YOUR UNDERSTANDING 1.7

The runner's speed is 8.3 meters per second.

CHECK YOUR UNDERSTANDING 1.8

The temperature is 77°F and 25°C.

CHECK YOUR UNDERSTANDING 1.9

There are 525,960 minutes in a year.

CHECK YOUR UNDERSTANDING 1.10

The temperature at 9 A.M. was 3°C.

CHECK YOUR UNDERSTANDING 1.11

No snow fell during December 19 and December 23.

CHECK YOUR UNDERSTANDING 1.12

Half of the precipitation events were rain.

CHAPTER 2 EXPERIMENTAL DESIGN

BIG IDEA CHECK-UP

1. **C.** Graduated cylinders are used to measure liquids. A ruler would be used to measure the length of the stems. A thermometer measures temperature. A calendar would be used to keep track of time.

2. **D.** In order to discover the effect of water, it must be compared to a condition without water.

3. **The following answer would receive 3 points because it identifies the best result and states the conditions.**

   Group 2 had the most plants and the biggest plants. The seeds need 2 ml of water every other day.

The following answer would receive 2 points because it states the best condition, but does not explain why it is the best.

The seeds need 2 ml of water every other day because without water the seeds died and with too much water the plants died.

The following answers would receive 1 point because they state a condition, but have no explanation.

The seeds need 2 ml of water every other day.

or

The seeds will die with too much water or not enough water.

The following answer would receive 0 points because although an answer was attempted it is incomplete or off-topic.

Plants need water.

CHAPTER 3 LIFE SCIENCE

BIG IDEA CHECK-UP 3.1

1. **C.** Compost is the answer because it is no longer like the log. The log is dead, which is different from nonliving. The tree is alive. The mushroom is alive.

2. **B.** Livings things are capable of reproduction.

BIG IDEA CHECK-UP 3.2

1. **D.** The Sun is the source of energy for life on Earth.

2. **B.** The oak tree is living. The wooden park bench is nonliving.

3. **B.** Flowers are plants. The insect and the bird are consumers. The Sun is an energy source.

4.　　sun ➜ flower ➜ insect ➜ bird ➜ snake

5.　　**The following answer would receive 3 points because it gives the correct answer and an explanation.**

The snakes will eat birds. Birds eat insects. If there are fewer birds, there will be more insects. The predators are reduced.

The following answer would receive 2 points because it is a correct answer but the explanation is incomplete.

There will be more insects because there are fewer birds.

The following answers would receive 1 point because it is correct, but has no explanation or an incorrect explanation.

There will be more insects.

or

There will be more insects because there are snakes in the food chain.

The following answer would receive 0 points because although an answer was attempted it is incomplete or off-topic.

The snake eats the bird.

BIG IDEA CHECK-UP 3.3

1. **D.** Plants and animals both need water. Plants need soil and carbon dioxide. Animals need oxygen.

2. **B.** Animals exhale carbon dioxide.

3. **The following answer would receive 1 point for a correct relationship.**

   Animals need plants for food. Animals need plants for oxygen.

BIG IDEA CHECK-UP 3.4

1. **B.** Leaves contain chloroplasts needed for photosynthesis.

2. **A.** The plant would not be able to make food because there would be no light for photosynthesis.

3. **D.** Sunlight is needed as the source of energy for the system. The jar will trap heat. The plant will provide oxygen. The worm will provide carbon dioxide. The plant will make its own food. The worm will obtain food from organic matter in the soil.

4. The following answer would receive 3 points because it is correct and explains the relationship between the roles of the honeybee and what happens when there are fewer bees.

    Apples come from flowers. Honeybees pollinate flowers. If there are fewer honeybees, some trees may not be pollinated. There would be fewer apples.

    The following answer would receive 2 points because it is incomplete.

    Fewer trees would be pollinated because there are not enough bees.

    or

    There will be fewer apples because fewer trees are pollinated.

    The following answer would receive 1 point because there is no explanation.

    There will be fewer apples.

    The following answer would receive 0 points because although an answer was attempted it is incomplete or off-topic.

    Bees pollinate flowers.

BIG IDEA CHECK-UP 3.5

1. **D.** The shell helps protect the snail.

2. **A.** Earthworms breathe through their skin. They could drown if the soil is too wet.

BIG IDEA CHECK-UP 3.6

1. **C.** The nervous system controls body systems.

2. **D.** Smell enhances taste.

3. **D.** Breathing is controlled by the brain.

4. The following answer would receive 3 points because it explains the need to breathe faster, the role of the heart, and why the rate increases and decreases.

    When you run, you need more oxygen. You need to breathe faster. Your heart needs to beat faster to move oxygen to your muscles. When you stop running, you don't need as much oxygen.

The following answer would receive 2 points because it does not explain the feedback between systems.

You need more air when you run so you breathe faster and your heart beats faster. You don't need as much air when you rest.

The following answers would receive 1 point because only one condition is discussed.

You need more air when you run.

or

You need less air when you are standing.

The following answer would receive 0 points because although an answer was attempted it is incomplete or off-topic.

You need to breathe when you run.

BIG IDEA CHECK-UP 3.7

1. **B.** Birds are the only animals that have feathers.

2. **C.** The crab and lobster both have ten legs and a hard outer skeleton. A bat is an animal with hair or fur. Birds have feathers. Spiders have eight legs. Grasshoppers have six legs. Dolphins are animals that have hair or fur. Fish have scales.

BIG IDEA CHECK-UP 3.8

1. **B.** The bird with the small beak would have the advantage. Bird A has a beak adapted for cracking hard nuts. Bird C has a beak adapted for collecting nectar from flowers. Bird D has a bill adapted for eating fruits.

BIG IDEA CHECK-UP 3.9

1. **A.** Grasshoppers are insects that go through incomplete change.

2. **C.** Embryos 2 and 4 have the most number of similarities.

CHAPTER 4 PHYSICAL SCIENCE: CHEMISTRY

BIG IDEA CHECK-UP 4.1

1. **A.** If the balloon with air has more mass than the balloon without air, then the difference is the mass of air. Answer B shows that air is less dense than water. Answer C shows that air takes up space. Answer D demonstrates that gas molecules have less energy when they are cold.

2. **C.** Using a magnet takes advantage of a property that iron has but that sand does not have. The iron filings will be attracted to the magnet. Answer A, tweezers, would not be easy, and it may not be possible to completely separate the two substances. Answer B may not work if the filings and sand are about the same size. Answer D, a magnifying glass, would allow you to see the two substances, but it would not separate them.

3. **A.** A property of gases is that they expand as the temperature increases and shrink when chilled. Answers B and C are incorrect because expansion is not a property of solids and liquids. Answer D is incorrect because a vapor is a substance that has formed a gas below its boiling point. Helium is a gas at normal temperatures, not a vapor.

BIG IDEA CHECK-UP 4.2

1. **B.** Water is a liquid between 32 degrees Fahrenheit and 212 degrees Fahrenheit.

2. **C.** Liquids will take the shape of the container into which they are poured, but they occupy a definite amount of space. Liquids cannot be forced into a smaller space.

BIG IDEA CHECK-UP 4.3

1. **C.** A change of phase is a physical change. The amount of water did not change. The mass will be the same.

2. **B.** The volume of the ice cube has already displaced the liquid water. What about the small part of the ice cube above the water line? Because ice expands when frozen it takes up less space when it thaws. When ice cubes melt in water, they do not cause an increase in the water level.

BIG IDEA CHECK-UP 4.5

1. The following answer would receive 3 points because it answers the question and uses evidence to support a conclusion.

A chemical reaction took place between vinegar and baking soda. The color was more like water than vinegar. A new substance was formed.

The following answers would receive 2 points because they answer the question and provide either evidence or a conclusion.

A chemical reaction took place. A new substance formed.

or

A chemical reaction took place. The color changed.

The following answers would receive 1 point because they only gave an answer, evidence, or a conclusion.

A chemical reaction took place.

or

A new substance formed.

The following answer would receive a 0 because it describes what happened but not why.

The color changed.

2. The following answer would receive 3 points because it answers the question and uses evidence to support a conclusion.

    No chemical reaction took place between water and salt. The color was the same as water. No new substance was formed.

    The following answers would receive 2 points because they answer the question and provide either evidence or a conclusion.

    No chemical reaction took place between water and salt. The color was the same as water.

    or

    No chemical reaction took place between water and salt. No new substance was formed.

    The following answers would receive 1 point because they only gave an answer, evidence, or a conclusion.

    No chemical reaction took place.

    or

    No new substance formed.

    The following answer would receive a 0 because it describes what happened but not why.

    The color did not change.

CHAPTER 5 PHYSICAL SCIENCE: PHYSICS

BIG IDEA CHECK-UP 5.1

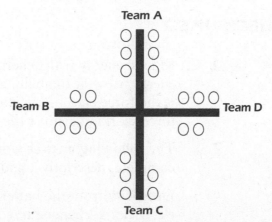

1. **A.** Team A has one extra student. This will cause an unbalanced force. The flag will move toward Team A.

2. **B.** Mark was able to run 1 mile in the least amount of time. He had the fastest speed.

3. The following answer would receive 3 points because it states what will happen, provides details about the motion, and explains what caused the motion.

When Magnet #1 is moved closer to Magnet #2, Magnet #2 will move away from Magnet #1. Magnets have poles. North poles push against each other.

The following answers would receive 2 points because they state what will happen but are missing either details or an explanation.

Magnet #2 will move away from Magnet #1 because magnets have poles.

or

Magnet #2 and Magnet #1 will move away from each other. North poles push against each other.

The following answers would receive 1 point because only the motion or the reason is given.

North poles push against each other.

or

Magnet # 2 will move away from Magnet #1.

The following answer would receive 0 points because it is incomplete and unclear.

It will move.

BIG IDEA CHECK-UP 5.2

1. **C.** Kinetic energy is the energy of motion. Answer A, potential energy, is what the ball had before it fell. Answer B, gravity, is the force causing the motion. Answer D, friction, is the force working against the motion.

2. The following answer would receive 3 points because it is complete, descriptive, and correct.

Electricity from the battery goes through the wires. The electricity is changed to sound energy by the bell.

The following answer would receive 2 points because it lacks detail.

The battery makes electricity. The bell makes sound.

The following would receive 1 point because it does not state the energies that are involved in the transformation.

The battery makes the bell ring.

The following would receive 0 points because it is off-topic.

When you push the switch the bell rings.

BIG IDEA CHECK-UP 5.3

1. **A.** Rubber band A would have the lowest pitch because all the bands are the same length, but A is the thickest.

2. The following answer would receive 3 points because it is correct, describes what would happen, and gives a specific explanation.

   The sand will move because the rubber bands make sound. Sound causes vibrations. Vibrations will move the sand.

   The following answers would receive 2 points because they are correct, but the explanations need to be more specific.

   The sand will move because of the sound.

   or

   The sand will move because of vibrations.

   The following answer would receive 1 point because it does not provide an explanation.

   The sand will move.

   The following answer would receive 0 points because it is off-topic.

   The rubber bands will make noise.

BIG IDEA CHECK-UP 5.4

1. **D.** Heat is transferred from an object that is at a higher temperature to one at a lower temperature. The air was the same temperature as the water. The ice was colder than the water. The water gave up heat to the ice.

2. B. After the ice melted, the water was colder than the air. The air transferred heat to the water.

3. There are many possible solutions to this problem, but they all involve blocking the transfer of heat from the air to the water in the glass.

An answer would receive 3 points if it described a condition with transfer of heat that would keep the water cool and was correct and detailed to show understanding.

Kevin could put the glass of water in the refrigerator. The water would give heat to the air in the refrigerator and stay cool longer.

or

Kevin could put the glass of water in a cooler. The cooler would keep the air from giving the water heat energy. It would stay cool longer.

or

Kevin could put the glass of water in a bowl of ice water. The water in the glass would give heat to the ice water. It would stay cool longer.

An answer that addressed the heat transfer without a flow description would receive 2 points.

Kevin could put the glass of water in a cooler to keep it cold. Things stay cold in a cooler.

An answer that showed an understanding that the glass needed to be somewhere cooler would receive 1 point.

Kevin could put the water in the refrigerator.

An answer that would earn 0 points would be off-topic.

Ice keeps water cold.

BIG IDEA CHECK-UP 5.5

1. **C.** Water allows light to pass through it. The others reflect light.

2. The following answer would receive 3 points because it is correct, describes why the shadow forms using the spatial relationship between the Sun, tree, and grass, and describes interaction of light with an opaque object.

   The tree is between the Sun and the grass. Because the sunlight cannot go through the tree, we see a shadow.

   The following answers would receive a 2 because they give the reason, but they do not describe the spatial relationship that causes the shadow.

   The tree is opaque.

   or

   The tree blocks the light.

   The following answer would receive 1 point because it is very general.

   Shadows mean that light is stopped by something.

   The following would receive 0 points because it is off-topic.

   Shadows happen when the sun is shining.

BIG IDEA CHECK-UP 5.6

1. **D.** The battery turns chemical energy into electrical energy. The wires carry the electricity. The switch controls the flow of electricity. The light bulb uses electricity.

2. **B.** The battery is a chemical reaction that makes electricity that can be turned into light energy.

CHAPTER 6 EARTH SCIENCE

BIG IDEA CHECK-UP 6.1.A

1. **B.** Because metamorphic rocks are formed by high pressure they tend to form bands of similar materials. Rock a and c are sedimentary rocks. Rock d is an igneous rock.

2.   A fourth grader would not be expected to know about sea floor spreading, but should be able to understand two ways that igneous rocks can form.

The following answer would receive 3 points, because it is correct, demonstrates knowledge of how igneous rocks form, and applies the knowledge to an event.

Lava or magma hardens to form igneous rocks. The Great Falls are igneous rock. There must have been a volcano or something that let magma come to the surface of the Earth 200 million years ago.

The following answers would receive 2 points, because they are correct, but only mention the formation of igneous rocks or an event.

Igneous rocks come from magma and lava.

or

Igneous rocks come from volcanoes.

The following answer would receive 1 point, because it is a possible conclusion, but it is unclear how the conclusion was reached.

Maybe a volcano was there.

The following answer would receive 0 points because it is off topic.

Water can turn igneous rocks into sediments.

BIG IDEA CHECK-UP 6.1.B

1. **D.** Erosion breaks rocks into sediments. Crystallization brings chemicals together. Pressure creates a metamorphic rock. Melting is the beginning of an igneous rock.

2. **A.** Metamorphic rock forms in folded faults from the pressure.

BIG IDEA CHECK-UP 6.1.C

1. **A.** Mineral A is attracted to a magnet.

2. **B.** Mineral B is able to scratch all the other minerals. Therefore it is the hardest.

BIG IDEA CHECK-UP 6.1.D

The following answer would receive 3 points, because it defines what soil is, identifies the brown material, and applies the definition to the brown material. The brown material could be identified as organic matter or decomposed living things, or compost.

Soil is a mixture of minerals, sediments, and organic matter. The brown material is organic matter. Organic matter is part of soil, but not soil.

The following answer would receive 2 points, because it is missing a definition, identification, or application of the definition.

The compost is not soil, because it is only organic matter.

or

Soil is partly organic matter, so the brown material is not soil.

The following answer would receive 1 point, because how the conclusion was reached is unclear.

The brown material is not soil.

The following answer would receive 0 points because it is off-topic.

The brown material is compost.

BIG IDEA CHECK-UP 6.1.E

1. **C.** Clay is very tightly packed. It does not allow water to flow through it easily.

2. **D.** Water runs quickly through D. It is probably sand. Sand is a good cactus garden soil.

3. The following answer would receive 3 points, because it discusses what each of the three layers does.

The sand and compost give plants minerals for growth. The sand layer would let water drain so that the soil is not too wet. The gravel would store water so that the soil does not dry out too much.

An answer that only discusses 2 parts of the garden would receive 2 points.

The top layer is good for plants. The bottom layer drains water.

An answer that is general would receive 1 point.

Water would drain into the gravel.

The following answer would receive 0 points because it is off-topic.

Gardens are a good place to grow plants.

BIG IDEA CHECK-UP 6.2

1. A. The raised shell formed when sediments filled a mold. The result is a cast.

2. C. Fossils form in sedimentary rock. The processes that form igneous and metamorphic rocks usually destroy fossils. A mineralized rock would be a petrified fossil.

3. B. The fossil resembles modern day scallops that live in the ocean. We apply what we know of modern animals to ncient animals.

BIG IDEA CHECK-UP 6.3

1. A. Nitrogen is more than three-quarters of the graph.

BIG IDEA CHECK-UP 6.4

1. D. Evaporation removes water from the soil. Condensation, precipitation, and runoff all add water to the soil.

2. B. The Sun is the main source of energy for Earth.

BIG IDEA CHECK-UP 6.5

1. C. Psychrometers measure relative humidity by determining how much more water the air can hold. Anemometers measure wind speed. Barometers measure air pressure. Thermometers measure temperature.

2. **A.** Expect rain in southern New Jersey, a mix of rain and snow in central New Jersey, snow in northern New Jersey. The snow is approaching from the west and will move across northern NJ. The warm wet air pushing north will keep the colder air to the north. The temperatures determine the precipitation types.

3. **C.** Ocean temperatures are warmest in August–September.

4. **A.** Storm surges occur where the ocean meets land. They cause a great deal of damage to beaches and coastal communities.

5. **C.** The lamp is what is different between the two experiments.

6. **A.** As the hot air rises and the cold air sinks the air molecules bump into the thermometer.

7. **D.** The air left the bottom of the bottle when the lamp was turned on and started to heat the air.

8. **A.** The relative humidity would not change if there was no water.

9. **C.** Water molecules need dust particles to collect on to form a cloud.

BIG IDEA CHECK-UP 6.6

1. **D.** Weathering is a slow process. The others cause sudden changes.

2. **B.** Volcanic eruptions formed the Watchung Mountains.

3. **C.** Glaciers covered the area 14,000 years ago.

4. **D.** Headwaters are the beginning of a river.

5. **C.** Sedimentary rocks are not as hard as igneous or metamorphic rocks.

6. **B.** Freezing and thawing are the most likely cause.

BIG IDEA CHECK-UP 6.7

1. **D.** Satellite images allow meteorologists to watch the hurricane as it moves.

2. **B.** A local map would provide the detail needed for planning the community.

CHAPTER 7 ASTRONOMY AND SPACE SCIENCE

BIG IDEA CHECK-UP 7.1

1. **D.** Of the planet choices, Uranus is the farthest from the Sun.

2. **B.** Venus and Mars are both planets.

BIG IDEA CHECK-UP 7.2

1. **D.**

Light from the Sun would strike the side of the Earth that is facing the Sun. The other side would be in darkness.

BIG IDEA CHECK-UP 7.3

1. **D.** The first day of winter has the least number of daylight hours in the northern hemisphere because the Earth is tilted away from the Sun.

2. **A.** This experiment proves that the Earth is tilted on its axis. The change in shadow length shows the change in angle between the Sun and Earth during the orbit. If the Earth were upright on its axis, the shadow would not change as the Earth orbited the Sun. The shadow changes because Earth is tilted on its axis.

BIG IDEA CHECK-UP 7.4

1. **C.** The light is moving from right to left decreasing the shadow on the right side.

2. **B.** The changes in phase take place in a 28-day period. The other answers are not correct for the following reasons. The Moon does complete one rotation on its axis every 28 days, but we do not see that from Earth. The Moon does not produce light. It reflects sunlight. If the moon disappeared into Earth's shadow that would be an eclipse, not a moon phase.

BIG IDEA CHECK-UP 7.5

1. **C.** Constellations are groups of stars. Stars are outside the solar system.

2. **C.** The Sun is a star and from Earth can only be seen during the day.

CHAPTER 8 ENVIRONMENTAL SCIENCE

BIG IDEAS CHECK-UP CHAPTER 8

1. **B.** A tree could be ready for harvesting in 80 years. Oil, coal, and ore would take far longer than 100 years to form.

2. **B.** Soybeans are a renewable resource. The other actions waste resources.

3. **D.** Electricity from coal energy uses a resource that cannot easily be replaced. Solar, wind, and tidal energy are considered perpetual energy sources or, in other words, something that doesn't run out.

4. To receive full credit for this answer, you would need to provide two examples, a reason why it would reduce trash, and a reason why it conserves resources.

   The examples should address recycling, or reduction of use, or reuse of materials. Not all fourth graders would be expected to know that juice boxes are recyclable.

   Materials for recycling would include bottles, cans, and juice boxes.

   Materials for reduction would be things that have reusable substitutes.

   Materials for reuse would be using the material again or turning it into a new use.

   The following are examples. Two of the following would need to be stated to receive 3 points.

   Recycling bottles and cans would save nine bags of trash each week. This would save aluminum and oil resources.

   Recycling juice boxes would save four bags of trash. This would save all the materials used to make the juice box.

Composting the food waste would save five bags of trash. This would save space at the landfill and reuse the materials for plants.

Using reusable sandwich containers would reduce plastic bag trash. This would save the oil needed to make plastic bags.

Using forks and spoons that can be washed and reused would reduce the plastic spoons and forks in the trash. This would save the materials used to make plastic.

Clean yogurt cups can be used as paint pans or for plant cups in science class. This would take them out of the trash and give them one more use before they become trash.

Answers that use similar ideas but do not explain how they reduce trash or save resources would receive 2 points.

Answers that use similar ideas with an explanation would receive 1 point.

Answers that are off-topic would receive 0 points.

INTRODUCTION TO PRACTICE TESTS

These practice tests are designed to mirror the actual NJ ASK4 Science tests you will take this spring. Here are a few helpful hints to help you succeed.

The Science test is usually administered on the last testing day of the week. It consists of four 15-minute sections each containing ten multiple-choice questions and one open-ended question. It is recommended that you make every effort to complete the multiple-choice questions within the first ten minutes and leave yourself a full five minutes for the open-ended question.

Read each question **carefully**. Often the questions are worded in such a way that incorrect answers are disguised by reversing the correct answer. Use special caution when the words "always" and "never" appear. Whenever possible try to narrow your choices to two possible correct answers.

Remember that the open-ended questions are graded using a rubric. Don't assume that the scorer knows what you mean (as your classroom teacher might!). Write everything out. Whenever possible use diagrams, tables, or pictures to support your answer and label **everything**!

After you have finished:

- Check all your answers or answer items you may have left out.

- Be sure that you marked the letter you intended for each.

- Make sure all open-ended questions are answered as completely and accurately as possible.

- Proofread your open-ended answers for errors.

- Review test directions to make certain you have completed everything required.

By knowing what to expect and practicing your skills in advance, you will put yourself in the best possible position to succeed on the NJ ASK 4 in Science.

NJ ASK4 SCIENCE PRACTICE TEST 1

DIRECTIONS FOR STUDENTS

Today you will take a Science test.

When you are taking this test, remember these important things:

1. Read each question carefully and think about the answer.
2. If you do not know the answer to a question, go on to the next question. You may come back to the skipped question later if you have time.
3. If you finish a section of the test early, you may check your work in that section only.
4. When you see a STOP sign, do **not** turn the page until you are told to do so.

SECTION 1

1. In the scientific method a **hypothesis** is
   A. something that can change.
   B. a decision made based on data.
   C. a possible answer to your question.
   D. an organized record of data.

2. A meteorologist would use a barometer to
   A. measure air pressure.
   B. measure temperature.
   C. see which way the wind is blowing.
   D. analyze data.

171

3. Through which of these do flying insects help plants reproduce?

    A. Photosynthesis

    B. Pollination

    C. Grafting

    D. Absorption

4. How are a daisy and a rabbit alike?

    A. They both make their own food.

    B. They both can move.

    C. They both eat other living things.

    D. They are both made of cells.

5. This part of a plant cell traps the Sun's energy to make food.

    A. Nucleus

    B. Chloroplast

    C. Cytoplasm

    D. Cell membrane

6. Two examples of animal instincts are

    A. reproduction and camouflage.

    B. migration and hibernation.

    C. classification and comparison.

    D. respiration and blinking.

7. The four things that plants need to live, grow, and reproduce are

    A. water, sunlight, soil, and CO_2.

    B. water, shelter, energy, and soil.

    C. water, oxygen, sunlight, and soil.

    D. water, sunlight, air, and food.

8. Which part of a flower becomes the fruit?

A. seed

B. petal

C. stamen

D. ovary

9. Which of the following is NOT true about the root system of a plant?

A. It anchors the plant in the ground.

B. It stores food for the plant.

C. It makes food for the plant.

D. It absorbs water and minerals from the soil.

10. The process by which plants make their own food is called

A. respiration.

B. fertilization.

C. photosynthesis.

D. chlorophyll.

OPEN-ENDED QUESTION

11. Jessica and Megan are cutting up a tomato. They examine the seeds and wonder if the tomato comes from a flowering plant. What would you tell them?

- Explain how you know the tomato comes from a flowering plant.
- Describe the stages in the life cycle of a flowering plant.

_____

_____

_____

_____

_____

 If you have time remaining of the 15 minutes allotted, you may review your work in this section only.

SECTION 2

In this section of the Science test, choose the correct answer for each multiple-choice question and then answer the open-ended question.

MULTIPLE CHOICE

12. The scientific name for all the living and nonliving things in an environment is

    A. habitat

    B. ecosystem

    C. population

    D. community

13. The name for the system by which organisms transfer energy by eating and being eaten is called

    A. predator/prey.

    B. parasite host.

    C. food chain.

    D. co-dependents.

14. In a simple food chain, a mouse eats grass and a snake eats the mouse. The primary role of the mouse in this food chain is to

    A. be a source of food energy.

    B. produce its own food.

    C. start a new species.

    D. decompose dead plant and animal parts.

15. When an environment and the communities in it change gradually over time from one community of organisms to another it is called

    A. climate.

    B. ecosystem.

    C. succession.

    D. evolution.

16. Climate change, human activities, and volcanic eruptions may all cause the dying out of a species.

A. Fossils

B. Succession

C. Extinction

D. Pollution

17. Three **non**living things that affect the organisms in an ecosystem include water, temperature, and

A. plants.

B. parasites.

C. soil.

D. decomposers.

18. The human organ system that supports and protects your internal organs and helps you to move is called the

A. digestive system.

B. respiratory system.

C. circulatory system.

D. skeletal system.

19. The main organ of the respiratory system that allows your body to obtain oxygen from the air you breathe is called

A. the heart.

B. the kidney.

C. the lung.

D. the diaphragm.

20. What is the name for the body system that helps protect your body from pathogens?

A. Central nervous system

B. Vaccination

C. Muscular system

D. Immune system

21. Which body structure links the brain to the rest of the human body?

    A. Esophagus

    B. Spinal cord

    C. Neuron

    D. Trachea

OPEN-ENDED QUESTION

22. Kristen is environmentally conscious. She tries to convince her brother John to buy a bicycle instead of a car because of the problems caused by fossil fuels and our need to conserve them. What should Kristen tell John?

    • Explain two problems caused by fossil fuels.
    • Give two reasons why conservation is worthwhile.

    _____

    _____

    _____

    _____

    _____

 If you have time remaining of the 15 minutes allotted, you may review your work in this section only.

SECTION 3

In this section of the Science test, choose the correct answer for each multiple-choice question and then answer the open-ended question.

MULTIPLE CHOICE

23. What percentage of the Earth's surface is covered with water?

    A. 23 percent

    B. 50 percent

    C. 60 percent

    D. 75 percent

24. The process of changing liquid water to water vapor is called

    A. precipitation.

    B. evaporation.

    C. boiling point.

    D. dew.

25. Air always moves in which of the following ways?

    A. From a place with high pressure to a place with low pressure.

    B. From a place with low pressure to a place with high pressure.

    C. From a cooler place to a warmer place.

    D. From over the ocean to over the land.

26. The amount of water vapor in the air determines its

    A. humidity.

    B. temperature.

    C. air pressure.

    D. density.

27. The sky is filled with thick, white, puffy cumulus clouds. What kind of weather should I expect to see?

    A. Thunderstorms

    B. Heavy rain

    C. Fair or good weather

    D. Fair followed by intense rain

28. Which of the following tools would not be used to measure or predict weather?

    A. Thermometer

    B. Scale

    C. Barometer

    D. Anemometer

29. Most of the Earth's fresh water is

    A. in the oceans.

    B. in clouds.

    C. underground.

    D. frozen in glaciers.

30. Which of the following statements comparing hurricanes and tornadoes is false?

    A. Hurricanes form over water, and tornadoes form over land.

    B. Both hurricanes and tornadoes have high winds.

    C. Both types of storms usually last only a few minutes.

    D. Both can cause great damage.

31. Which of the following is considered a mineral?

    A. Salt

    B. Diamond

    C. Granite

    D. All of the above

32. If a scientist is able to scratch a piece of calcite with a piece of topaz, what conclusion can he make about these two minerals?

   A. The calcite is harder than the topaz.

   B. The topaz has more luster than the calcite.

   C. The topaz is harder than the calcite.

   D. The topaz is bigger than the calcite.

OPEN-ENDED QUESTION

33. Kevin and Alison went hiking with a group of friends. They noticed a ledge made of layers of limestone. Nearby they see a piece of marble and a piece of obsidian. The students observe that these rocks represent the three ways that rocks can form. Do you agree with this observation? Why or why not?

   - Name the three categories of rocks.
   - Explain the processes by which each is formed.

   _____

   _____

   _____

   _____

 If you have time remaining of the 15 minutes allotted, you may review your work in this section only.

SECTION 4

In this section of the Science test, choose the correct answer for each multiple-choice question and then answer the open-ended question.

MULTIPLE CHOICE

34. All of the following are examples of landforms except
    A. plains.
    B. plateaus.
    C. habitats.
    D. canyons.

35. Which of the following statements is true?
    A. Volcanic eruptions can affect the climate.
    B. Tsunamis can cause earthquakes.
    C. Waves in the ocean cause erosion but not deposition.
    D. Wind causes most landslides.

36. Which of these are ways to make nonrenewable resources last longer?
    A. Energy efficient cars
    B. Recycling
    C. Turning off the lights when you leave a room
    D. All of the above

37. Anything that has mass and takes up space is called
    A. matter.
    B. solid.
    C. liquid.
    D. gas.

38. What is the best tool for comparing the mass of two types of matter?

A. Scale

B. Pan balance

C. Thermometer

D. Yardstick

39. When salt is mixed in water, which of the following statements is true?

A. Salt is the solute, and water is the solvent.

B. Water is the solute, and salt is the solvent.

C. The salt takes in, or dissolves the water.

D. There is usually more salt than water.

40. A chemical change occurs when

A. ice melts and becomes water.

B. wood burns to form ash.

C. clay is formed into a bowl.

D. salt is mixed with water.

41. What would happen if a balloon with a negative charge and a balloon with a positive charge are held up by strings next to each other?

A. They will move apart.

B. They will move toward each other.

C. Nothing will happen.

D. They will both fall to the floor.

42. When most liquids are heated to the boiling point, the tiny particles of matter that make up the liquid

A. become more closely packed together.

B. move more slowly.

C. move quickly in all directions.

D. take up a definite amount of space.

43. Aluminum foil and wood can be grouped together because of the way they react to light. They are classified as

   A. transparent.

   B. translucent.

   C. opaque.

   D. solid.

OPEN-ENDED QUESTION

44. Caitlyn's family is making dinner on a camping trip. Her brother Myles starts a campfire. Her sister Danielle melts butter in a frying pan. Myles tosses lettuce and tomatoes and makes an oil and vinegar salad dressing.

   • Which of these activities results in chemical changes and which in physical changes?
   • Identify any changes in phase that are occurring, and point out any mixtures.

   _____

   _____

   _____

   _____

   _____

 If you have time remaining of the 15 minutes allotted, you may review your work in this section only.

NJ ASK4 SCIENCE PRACTICE TEST 2

DIRECTIONS FOR STUDENTS

Today you will take a Science test.

When you are taking this test, remember these important things:

1. Read each question carefully and think about the answer.
2. If you do not know the answer to a question, go on to the next question. You may come back to the skipped question later if you have time.
3. If you finish a section of the test early, you may check your work in that section only.
4. When you see a STOP sign, do **not** turn the page until you are told to do so.

SECTION 1

1. The largest group that scientists can group living things into is a

   A. kingdom.

   B. family.

   C. species.

   D. class.

2. Ferns and mosses reproduce by forming tiny cells that can grow into new plants called

   A. seeds.

   B. spores.

   C. cones.

   D. needles.

183

3. Which of the following are vertebrates?

   A. Spiders and worms

   B. Jellyfish and clams

   C. Snakes and fish

   D. Crabs and lobsters

4. The four things that every animal needs to survive are food, oxygen, shelter, and

   A. fur.

   B. water.

   C. sunlight.

   D. soil.

5. A polar bear's white fur coat that helps it blend into its cold, snowy environment is an example of

   A. imprinting.

   B. mimicry.

   C. trait.

   D. camouflage.

6. Which of the following is an example of learned behavior?

   A. A bird flying south

   B. A garter snake warming in the sun

   C. A lion cub hunting

   D. A sea turtle returning home to reproduce

7. Which of the following is NOT true about an animal cell?

   A. It contains a gel-like liquid inside.

   B. It is surrounded by a cell wall.

   C. It contains a nucleus.

   D. It contains cytoplasm.

8. A radish is an example of a
   A. fibrous root system.
   B. taproot.
   C. root hair.
   D. stem.

9. This is the female organ of the flower where egg cells are produced.
   A. Stamen
   B. Sepal
   C. Petal
   D. Pistil

10. For a seed to form, this process must take place in a plant.
    A. Photosynthesis
    B. Fertilization
    C. Germination
    D. Irrigation

OPEN-ENDED QUESTION

11. Emilie is observing a one-celled organism through a microscope. She notices that the cell has a nucleus. Explain how Emilie can tell what kingdom the organism belongs to.

    • What else can Emilie tell about the organism by knowing its kingdom?
    • Draw and label the cell.

    _____

    _____

    _____

    _____

    _____

 **STOP** If you have time remaining of the 15 minutes allotted, you may review your work in this section only.

SECTION 2

In this section of the Science test, choose the correct answer for each multiple-choice question and then answer the open-ended question.

MULTIPLE CHOICE

12. The function of decomposers in the ecosystem is

   A. to break down waste products into minerals and nutrients.

   B. to release oxygen into the atmosphere.

   C. to produce food energy.

   D. to control the population of an ecosystem.

13. A flea lives on a dog using its blood supply to live and often making the dog sick. In this type of relationship, the flea is called

   A. a predator.

   B. a parasite.

   C. a host.

   D. a prey.

14. Oreo, a pet store guinea pig who has lived alone in a cage, is moved into a larger cage with two other guinea pigs. What must Oreo do to survive in his new environment?

   A. Oreo must hide in the cage.

   B. Oreo must build a new nest.

   C. Oreo must compete with the other two guinea pigs for the resources he needs.

   D. Oreo must wait to eat and drink until after the other two have finished.

15. What is the name for the large blood vessels that carry oxygen-rich blood from the left side of the heart to the rest of the body?

   A. Veins

   B. Arteries

   C. Capillaries

   D. Ventricles

16. Where does the process of digestion begin?

   A. Mouth

   B. Esophagus

   C. Stomach

   D. Small intestine

17. The control center of your body, which links all body systems and carries signals from one to another, is the

   A. brain.

   B. spinal cord.

   C. central nervous system.

   D. neuron.

18. What kind of weather can you expect to see when a warm air mass runs into a cold air mass to create a warm front?

   A. Thunderstorms

   B. Brief, heavy precipitation

   C. Steady, long-lasting precipitation

   D. Hail

19. An air mass is a huge body of air that has nearly the same

   A. temperature and humidity.

   B. ice and dust particles.

   C. temperature and clouds.

   D. humidity and precipitation.

20. A scientist who studies the weather is called

    A. a paleontologist.

    B. an anthropologist.

    C. a meteorologist.

    D. a geologist.

21. What is one possible effect of extra "greenhouse gases" in our atmosphere?

    A. Earthquake

    B. Increase in Earth's temperature

    C. Decrease in Earth's temperature

    D. Drought

OPEN-ENDED QUESTION

22. The table below records the volume of a balloon at a given temperature. What happens to the volume of the balloon as the temperature increases?

Temperature	17° C	22°C	27°C	32°C
Balloon volume	1,450 ml	1,475 ml	1,500 ml	1,525 ml

- What trend can you observe based on the data presented?
- What would you expect to happen should the temperature rise above 32°C? If it drops below 17°C?

_____

_____

_____

_____

_____

 If you have time remaining of the 15 minutes allotted, you may review your work in this section only.

SECTION 3

In this section of the Science test, choose the correct answer for each multiple-choice question and then answer the open-ended question.

MULTIPLE CHOICE

23. In science, work means
   A. doing a job.
   B. a push or pull on an object caused by another object.
   C. a force applied by a machine.
   D. what is done when a force moves an object.

24. How do scientists classify hurricanes?
   A. Amount of damage they do
   B. Temperature of the water
   C. Wind speed
   D. How long they last

25. Scientists identify minerals by testing all of these physical properties except
   A. color.
   B. luster.
   C. hardness.
   D. size.

26. Fossils are most likely to be found in which type of rock?
   A. Sedimentary
   B. Igneous
   C. Metamorphic
   D. Crystals

27. The gradual wearing away or changing of rock and soil caused by water, ice, temperature changes, wind, chemicals, or living things is called

    A. crystallization.

    B. weathering.

    C. sediment.

    D. calcification.

28. Which of the following are forces that drive the rock cycle?

    A. Heat

    B. Pressure

    C. Weathering

    D. All of the above

29. Lava that cools quickly on the Earth's surface forms igneous rocks that have

    A. large crystals.

    B. no crystals.

    C. very tiny crystals.

    D. crystals that settle in layers.

30. The object in our solar system with the strongest gravitational pull is

    A. Earth.

    B. the Moon.

    C. the Sun.

    D. Mercury.

31. Which of the following natural resources are nonrenewable?

    A. Solar energy

    B. Fossil fuels

    C. Water

    D. Soil

32. All of the energy that is stored in fossil fuels can be traced to

    A. swamps.

    B. offshore drilling.

    C. the Sun.

    D. humus.

OPEN-ENDED QUESTION

33. Identify a product in your household that would not have been in your parents' homes when they were growing up.

    • Name at least one positive effect of this product.
    • Name at least one negative effect of this product.

_____

_____

_____

_____

_____

_____

STOP If you have time remaining of the 15 minutes allotted, you may review your work in this section only.

SECTION 4

In this section of the Science test, choose the correct answer for each multiple-choice question and then answer the open-ended question.

MULTIPLE CHOICE

34. In which of the following states of matter are the particles farthest apart?

   A. They are all the same.

   B. Solid

   C. Liquid

   D. Gas

35. Three properties of matter that can be measured are

   A. size, shape, and color.

   B. mass, volume, and density.

   C. texture, smell, and taste.

   D. none of the above.

36. When a piece of paper is folded to form an origami swan

   A. a chemical change has occurred.

   B. a physical change has occurred.

   C. the particles of matter that make up the paper have changed.

   D. the type of matter has changed.

37. A chemical change results in a

   A. solution.

   B. phase change.

   C. different kind of matter.

   D. loss of energy or matter.

38. What does temperature measure?

   A. The amount of heat a material has

   B. · How fast the particles in matter are moving

   C. How much liquid is in the container

   D. The amount of matter

39. Which of the following would be the best conductor of heat energy?

   A. Wood

   B. Metal

   C. Marble

   D. Plastic

40. All thermal energy

   A. always moves from cooler areas to warmer areas.

   B. moves randomly from cool to warm and warm to cool.

   C. always moves from warmer areas to cooler areas.

   D. none of the above.

41. When two solids touch, heat energy is transferred by which of the following?

   A. Radiation

   B. Conduction

   C. Convection

   D. Currents

42. When studying sound energy, wavelength is

   A. the number of waves that pass a point in a certain amount of time.

   B. the part of the wave where particles are clustered together.

   C. the distance between a point on one wave and a similar point on another.

   D. a disturbance that moves energy through matter.

43. Pressing a key on a piano extra hard will change its

A. pitch.

B. frequency.

C. loudness.

D. none of the above.

OPEN-ENDED QUESTION

44. Look carefully at the illustration below.

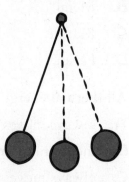

- On the illustration, identify the positions where the pendulum has the most kinetic energy and the positions where the pendulum has the most potential energy.
- Explain how the energy changes in the pendulum from potential to kinetic and back to potential.

_____

_____

_____

_____

_____

 If you have time remaining of the 15 minutes allotted, you may review your work in this section only.

ANSWERS TO PRACTICE TESTS

NJ ASK4 SCIENCE TEST 1

SECTION 1

1. **C.** A hypothesis is a possible answer to your question.

2. **A.** A meteorologist would use a barometer to measure air pressure.

3. **B.** Insects help plants reproduce through pollination.

4. **D.** A daisy and a rabbit are both made of cells.

5. **B.** The chloroplast is the part of a plant that traps the Sun's energy to make food.

6. **B.** Two examples of animal instincts are migration and hibernation.

7. **A.** The four things that plants need to live, grow, and reproduce are water, sunlight, soil, and CO_2.

8. **D.** The ovary of the flower becomes the fruit.

9. **C.** The roots do not make food for the plant.

10. **C.** Photosynthesis is the process by which plants make their own food.

11. **For an answer to receive full credit, it should contain the following elements:**

    I know that the tomato comes from a flowering plant because it has seeds. First, the tomato seed germinates, with the seeds absorbing water and the seed coat splitting open. Then, the plant uses the food inside the seed to grow, and the roots grow both down into the soil and up to the light. Leaves begin to make food for the plant, and the stem and leaves begin to grow. The plant grows into an adult plant and flowers begin to grow. The flower makes egg cells and pollen, pollination happens, and the fertilized eggs develop into seeds. Fruit forms around the seeds, the fruit and seeds separate from the plant, and the seeds germinate, beginning the cycle again.

SECTION 2

12. **B.** The scientific name for all the living and nonliving things in an environment is ecosystem.

13. **C.** The name for the system by which organisms transfer energy by eating and being eaten is called a food chain.

14. **A.** The primary role of the mouse in this food chain is to be a source of food energy.

15. **C.** Gradual change over time from one community of organisms to another is called succession.

16. **C.** The dying out of a species is called extinction.

17. **C.** Three nonliving things that affect the organisms in an ecosystem include water, temperature, and soil.

18. **D.** The skeletal system supports and protects your internal organs, and helps you to move.

19. **C.** The lungs are the main organ of the respiratory system.

20. **D.** The immune system helps protect your body from pathogens.

21. **B.** The spinal cord links the brain to the rest of the human body.

22.   **For an answer to receive full credit it should contain the following elements:**

    Two problems that fossil fuels cause are pollution and oil spills. Fossil fuel use causes pollution because particles from the burning of it end up in the air and make the air unhealthy to breathe. Carbon dioxide from the burning of fossil fuels may lead to global warming. Acid rain resulting from carbon dioxide can damage buildings and harm living things. Oil spills from oil drilling can cause pollution and can kill marine animals. Conservation is worthwhile because it helps the fossil fuels last longer and because it means less pollution from fossil fuels.

SECTION 3

23. **D.** 75% of the Earth's surface is covered with water.

24. **B.** The process of changing liquid water to water vapor is called evaporation.

25. **A.** Air always moves from a place with high pressure to a place with low pressure.

26. **A.** The amount of water vapor in the air determines its humidity.

27. **C.** When the sky is filled with thick, white, puffy cumulus clouds, we should expect fair or good weather.

28. **B.** Only the scale would not be used to measure or predict weather.

29. **D.** Most of the Earth's fresh water is frozen in glaciers.

30. **C.** While tornadoes may last for only a few minutes, hurricanes can last for days.

31. **D.** Salt, diamond, and granite are all minerals.

32. **C.** If a scientist is able to scratch a piece of calcite with a piece of topaz, the topaz is harder than the calcite.

33.   **For an answer to receive full credit it should contain the following elements:**

     The limestone ledge is sedimentary rock, the marble is metamorphic rock, and the obsidian is igneous rock. Metamorphic rocks are formed when rocks are changed by heat and pressure. Igneous rocks are formed when rocks melt and then cool. Sedimentary rocks are formed when layers of sediment form layers that harden. When your friend says that there are three ways that rock can form, he means that one kind of rock can melt and become igneous. Another kind of rock can be changed by heat and pressure and become metamorphic. A third kind of rock can break down, form sediments, and then harden into sedimentary rock.

SECTION 4

34. **C.** Plains, plateaus, and canyons are all examples of landforms. A habitat is not.

35. **A.** Volcanic eruptions can affect the climate.

36. **D.** Energy, efficient cars, recycling, and turning off the lights when you leave a room are all ways to make non-renewable energy sources last longer.

37. **A.** Anything that has mass and takes up space is called matter.

38. **B.** The best tool for comparing the mass of two types of matter is the pan balance.

39. **A.** When salt is mixed in water, salt is the solute and water is the solvent.

40. **B.** A chemical change occurs when wood burns to form ash.

41. **B.** If a balloon with a negative charge and a balloon with a positive charge are held up by strings next to each other, they will move toward each other.

42. **C.** When most liquids are heated to the boiling point, the tiny particles of matter that make up the liquid move quickly in all directions.

43. **C.** Aluminum foil and wood can be grouped together because light cannot pass through them. They are opaque.

44. **For an answer to receive full credit it should contain the following elements:**

    The physical change that is occurring is the melting butter. The campfire causes a chemical change in the materials it burns. The melting butter is a phase change, changing from a solid to a liquid. The lettuce and tomatoes combine to make a mixture. The oil and vinegar dressing is also a mixture.

NJ ASK4 SCIENCE PRACTICE TEST 2

SECTION 1

1. **A.** The largest group that scientists can group living things into is a kingdom.

2. **B.** Ferns and mosses reproduce by forming tiny cells that can grow into new plants called spores.

3. **C.** Snakes and fish are vertebrates.

4. **B.** The four things that every animal needs to survive are food, oxygen, shelter, and water.

5. **D.** A polar bear's white fur coat that helps it blend into its cold, snowy environment is an example of camouflage.

6. **C.** A lion cub hunting is an example of learned behavior.

7. **B.** An animal cell is NOT surrounded by a cell wall.

8. **B.** A radish is an example of a taproot.

9. **D.** The female organ of the flower where egg cells are produced is the pistil.

10. **B.** For a seed to form, fertilization must take place in a plant.

11. **For an answer to receive full credit, it should contain the following elements:**

    Emilie knows that two of the ways scientists tell what kingdom an organism belongs to are whether it is one-celled and whether it has a nucleus. Because this organism is one-celled and has a nucleus, it belongs in the protist kingdom. Because this organism is a protist, Emilie knows that it has other cell parts, but no tissue systems. Emilie also knows that this organism lives in either water or a moist environment, and that it either makes its own food or must get it. Emilie knows this organism may be algae, ameba, or paramecia.

SECTION 2

12. **A.** The function of decomposers in the ecosystem is to break down waste products into minerals and nutrients.

13. **B.** The flea is a parasite. The dog is the host in this relationship.

14. **C.** Oreo must compete with the other two guinea pigs for the resources he needs.

15. **B.** The large blood vessels that carry oxygen-rich blood from the left side of the heart to the rest of the body are the arteries.

16. **A.** The process of digestion begins in the mouth.

17. **C.** The control center of your body, which links all body systems and carries signals from one to another, is the central nervous system.

18. **C.** Expect to see steady, long-lasting precipitation when a warm air mass runs into a cold air mass to create a warm front.

19. **A.** An air mass is a huge body of air that has nearly the same temperature and humidity.

20. **C.** A scientist who studies the weather is called a meteorologist.

21. **B.** One possible effect of extra "greenhouse gases" in our atmosphere is that the Earth's temperature would increase.

22. **For an answer to receive full credit, it should contain the following elements:**

    As the temperature increases so does the volume of the balloon. The trend noticed is for each 5°C increase in temperature, the volume of the balloon increases by 25 ml. I would expect that if the temperature were to go above 32°C, the volume would be greater than 1,525 ml. However, at a certain temperature, I would expect that the balloon would pop. Likewise, if the temperature were to fall below 17°C, the volume of the balloon would fall below 1,450 ml with the balloon becoming almost completely deflated at some point.

SECTION 3

23. **D.** Work is what is done when a force moves an object.

24. **C.** Scientists classify hurricanes by wind speed.

25. **D.** Scientists identify minerals by testing color, luster, and hardness. Size is not tested.

26. **A.** Fossils are most likely to be found in sedimentary rock.

27. **B.** The gradual wearing away or changing of rock and soil caused by water, ice, temperature changes, wind, chemicals, or living things is called weathering.

28. **D.** Heat, pressure, and weathering are all forces that drive the rock cycle.

29. **C.** Lava that cools quickly on the Earth's surface forms igneous rocks that have small crystals.

30. **C.** The object in our solar system with the strongest gravitational pull is the Sun.

31. **B.** Fossil fuels are a nonrenewable natural resource.

32. **C.** All of the energy that is stored in fossil fuels can be traced to the Sun.

33. **For an answer to receive full credit, it should contain the following elements:**

In my household, we have cell phones for the whole family. These were not invented when my parents were growing up. A positive effect is that these help us stay in touch and let other family members know if someone is running late. A negative effect is that cell phone towers are ruining landscapes.

SECTION 4

34. **D.** Particles of matter are farthest apart in the gaseous state.

35. **B.** Three properties of matter that can be measured are mass, volume, and density.

36. **B.** When a piece of paper is folded to form an origami swan, a physical change has occurred.

37. **C.** A chemical change results in a different kind of matter.

38. **B.** Temperature measures how fast the particles in matter are moving.

39. **B.** Metal would be the best conductor of heat energy.

40. **C.** All thermal energy moves from warmer areas to cooler areas.

41. **B.** When two solids touch, heat energy is transferred by conduction.

42. **C.** Wavelength is the distance between a point on one wave and a similar point on another.

43. **C.** Pressing a key on a piano extra hard will change its loudness.

44.   **For an answer to receive full credit, it should contain the following elements:**

The pendulum in the two higher positions (left and right) should be labeled "potential energy" and the pendulum in the lower position (center) should be labeled "kinetic energy."

When the pendulum is at its highest position, it has the greatest amount of potential, or stored, energy. As the pendulum starts to swing down, the potential energy changes to kinetic energy, which is the energy of motion. At its lowest position on the path, the pendulum has the greatest kinetic energy. As the pendulum swings back up, the kinetic energy decreases and the potential energy increases.

INDEX

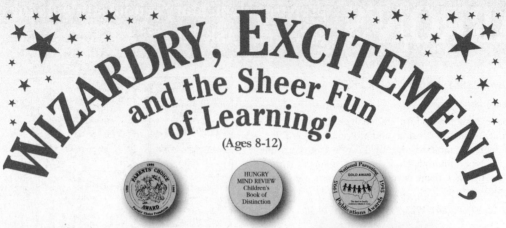

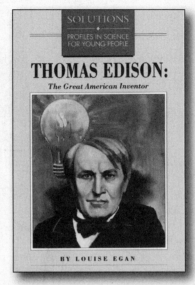
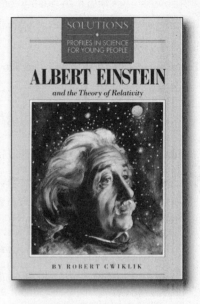